Lithographed Paper Toys, Books, & Games

1880-1915

Judith Anderson Drawe & Kathleen Bridge Greenstein

Schiffer Publishing Ltd
4880 Lower Valley Road, Atglen, PA 19310 USA

Dedication

This book is dedicated to Josh and Lindsey — the next generation of antique toy lovers. And to Edna Mae, who would be proud of her girls; Judy Mae and Kathy Mae.

With cries of joy, then, Frank and Will,
Soon had their boxes opened wide;
What store of precious toys were there,
With many useful things beside.
A piece of Fairy-land it seemed.
Or something, that the boys had dreamed.

The Three Boxes
McLoughlin Bros., 1888.

Our cover photo was taken from a poster entitled *Printed Pastimes: Nineteenth Century Lithographed Paper Toys*. This poster was issued in December, 1998 to commemorate the Linda and Carl Thomas collection of paper litho toys at the Brandywine Museum, Chadds Ford, Pennsylvania. The poster is available for purchase and can be obtained by e-mailing Linda and Carl Thomas (carl@caltiques.com).

Library of Congress Catalog Card Number: 99-69969

Designed by Bonnie M. Hensley
Type set in Seagull Hv BT/Aldine721 BT

ISBN: 0-7643-1124-7
Printed in China
1 2 3 4

Published by Schiffer Publishing Ltd.
4880 Lower Valley Road
Atglen, PA 19310
Phone: (610) 593-1777; Fax: (610) 593-2002
E-mail: Schifferbk@aol.com
Please visit our web site catalog at **www.schifferbooks.com**

In Europe, Schiffer books are distributed by Bushwood Books
6 Marksbury Avenue Kew Gardens
Surrey TW9 4JF England
Phone: 44 (0)208-392-8585; Fax: 44 (0)208-392-9876
E-mail: Bushwd@aol.com
Free postage in the UK., Europe; air mail at cost.

This book may be purchased from the publisher.
Include $3.95 for shipping. Please try your bookstore first.
We are always looking for authors to write books on new and related subjects.
If you have an idea for a book please contact us at the address above.
You may write for a free printed catalog.

Contents

THE
LITTLE POSTCARD PAINTERS

HERE
and
THERE

YACHT RACE

The OLD WOMAN
And her
PIG.

Acknowledgments

Many thanks to the following people for their help in bringing to life this beautiful book. To Paul Fink for letting us photograph his wonderful game collection. To Linda and Carl Thomas for their hospitality and for letting us view the most exquisite paper litho toy collection we've ever had the honor to behold. To Nori Cerny and Michelle Otremba for traveling so far to help a friend in need. To Shirley Chilberg and Jean Champion for not complaining when they were dragged out on countless antique hunts. To Linda Boltrek, Judy McClymond, Cathy Cook, and Austern's Antiques (Bellmore, New York) for contributing their treasures. To Donna Baker, our editor, for being so helpful and positive from day one. And last but not least, to Marc and John for being so wonderful and supportive during the whole process of making this book, we couldn't ask for better husbands!

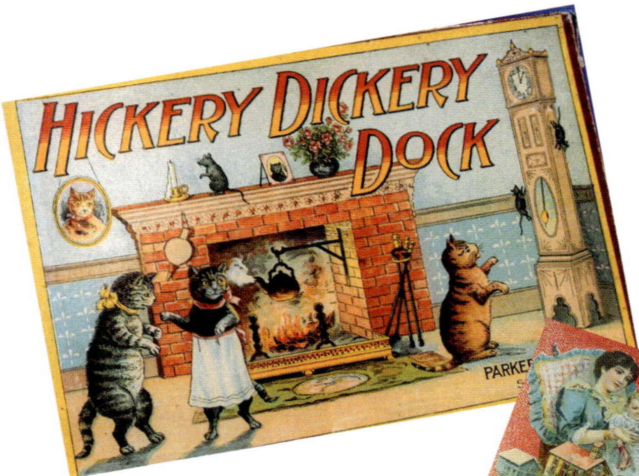

Introduction

In the late nineteenth century the hardships of the early part of the century gave way to increasing prosperity and security. Although life for children at the end of the 1800s was still very simple, there was a more relaxed moral and religious code. The entertainment of children centered around the home; radio, television and the movie theater did not exist for these children, so they had to depend on their own ingenuity when it came to playtime. Children played outdoor games like hide and seek and blind man's bluff or hoops. "House games" consisted of backgammon, checkers, jack straws, and fox and geese. Cashing in on this prosperity, many printing companies decided to not only attract children with their beautifully decorated toys and books, but also the parents — who controlled the purse strings. Instead of using television commercials to hook their audience, late nineteenth century toy manufacturers relied on the ornate and beautifully illustrated paper that was applied to their toys to sell their wares.

Around 1820, chromolithography (or color printing on paper) was invented in Germany. Although not widely used until the 1880s, it heralded a new way in which children's toys and games were produced. Lithography is a process in which a grease crayon is used to draw images on a smooth piece of limestone. The crayon section attracts the ink and repels water. Prior to chromolithography all children's toys had to be hand painted. This was a very time consuming and expensive procedure and did not produce the same depth and brightness of color as chromolithography. Chromolithography involved using multiple plates for each color in the print. These color plates had to be perfectly aligned on top of the lithographed print to insure that the different colors would not run into each other. What is remarkable about chromolithography is the fact that only one color could be used at a time. It is very easy to appreciate the artistry and the time this process took. Even though chromolithography was faster than hand tinting it was still an amazing artistic process.

Many of the early artists who produced the beautiful chromolithography at the end of the nineteenth century were never given the credit they so richly deserved. Artists like Palmer Cox, Thomas Nast, and Howard Pyle were only a few of the world's best-known artists who produced exquisite color lithographed pictures. Many others would go unknown but not unappreciated by collectors of paper lithographed toys.

The outstanding chromolithography in early children's toys

gives us a glimpse into another time, a time when families got together every Sunday night in their parlors and played games together. The chromolithography on late Victorian era toys pictured historical personages, important events and discoveries, new modes of transportation, holidays, sports, children, adventure, geography, animals, nursery rhymes, wars, finance, and patriotism.

It is interesting for the collector to note the depiction of minorities in early paper lithographed toys. Black and immigrant images were usually portrayed in very stereotypical and derogatory fashion. This type of memorabilia is sometimes collected to remind us of how far we have come in the last one hundred years.

Most of the paper lithographed toys in this book are at least one hundred years old. Although finding paper litho toys in mint condition is a very big and expensive undertaking, many new specimens are still being discovered every day, tucked away in an attic (the infant mortality rate was high at the turn of the twentieth century so many toys were put away having never been used by loving family members) or preserved by a family for nostalgia's sake. We also hope you will not turn away that book or set of blocks that is yellowed with age or that game that is missing a small piece of its apron. Just like an old Steiff bear that has lost part of its ear, we feel this is part of the ambiance of a time gone by that makes paper litho toy collecting an enchanting and affordable pastime.

This book celebrates the exquisite artistry that produced the brilliant chromolithography on paper found in children's toys from the late 1800s until the early twentieth century. As collectors and lovers of a time long past we wanted to show you the colorful world of Victorian paper litho toys. These are collectibles that are still affordable to the beginning collector and at the same time can be viewed in many prestigious museums around the world.

Pricing

The values in this book are meant to be a guide for the beginning collector. The value of any antique is based on rarity and condition. The first value listed is for a toy in good condition. The second value listed is for a toy in excellent condition with all original pieces.

Chapter One

Games

The first producers of games were originally publishers of maps or, as in the case of the McLoughlin Brothers, publishers of books. Until the mid-nineteenth century, Americans imported a lot of their games from abroad. Early board game producers looked to England and Germany for game ideas and strategies. Some of the first board games, like "The Mansion of Happiness" by W. & S.B. Ives in 1843 and "The Checkered Game of Life" by Milton Bradley in 1860, stressed the importance of honor, truth, wealth, ambition, and bravery as well as the ills of gambling, idleness, disgrace, and ruin. Later games stressed learning through play.

With the advent of the industrial revolution, people had more leisure time and more disposable income. Along with the arrival of the industrial revolution came the widespread use of chromolithography in printing. The "Kings of Chromolithography" were the McLoughlin Brothers of New York. Their games, books, and toys outshone many of their competitors and are still a marvel to see. Many other companies, like Milton Bradley and Parker Brothers also produced beautiful games. Germany was an important exporter of many games, producing them in many languages. Some European toy makers had printing shops in Germany. Spears & Sons of London, for example, had their own printing facility in Germany. Selchow and Righter of New York City were toy jobbers. They would buy and import games from different manufacturers and then sell them to the public in America.

The first board games consisted of a four fold board accompanied by a small box that would hold the implements needed to play the game. In most cases, these implements would include metal figures and some kind of spinner or teetotem for moving around the board. Dice were considered a gambling device by moralists and not used very often in early board games.

Parlor and skill games were very popular in the late 1800s. Families got together once a week and played games where difficulty and skill were less important than the togetherness the games inspired. Table top yacht races and miniature croquet were a few games you could play together no matter what the weather.

Small boxed games were very popular at the turn of the twentieth century and were used to teach learning and at the same time to make learning fun. There were games of travel, geography, conversation, and history.

Game of War, unmarked, c. 1870. Folding game board written in French, German, and English. 14" x 12". $100-150.

Boxed card games boasted beautifully lithographed pictures, the most popular being Snap, Logomachy, and of course Old Maid. In addition to games of instruction, there were games called "guesses," which included riddles of various kinds. In 1843, W. & S.B. Ives of Salem, Massachusetts, published Dr. Busby. These lithographed cards were first colored by hand and then later reproduced in chromolithography.

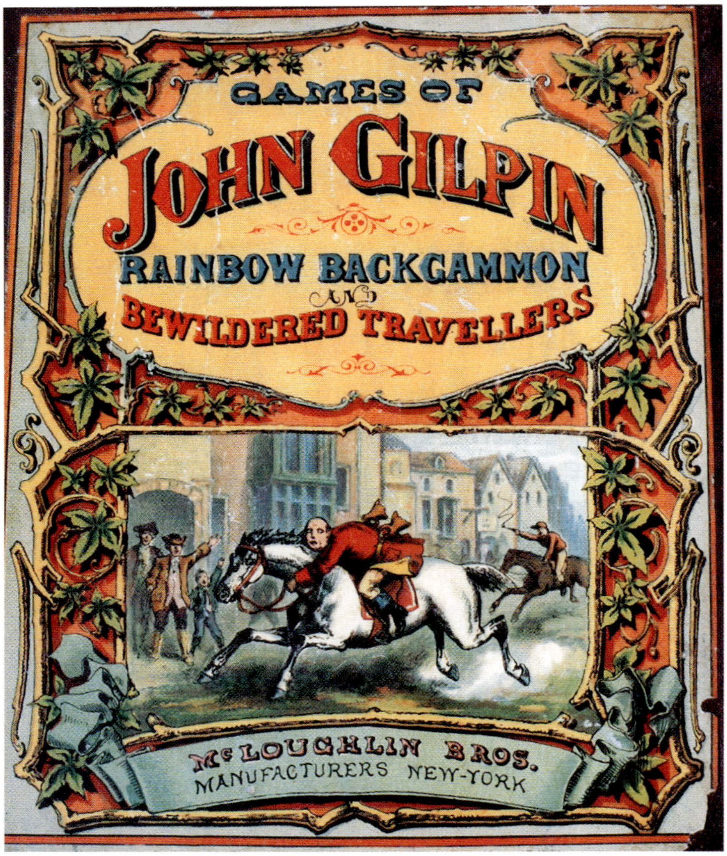

Games of John Gilpin, Rainbow Backgammon and Bewildered Travelers, By McLoughlin Bros., 1870. Bookcase type game with paper litho on wood spinner. Opens to 18" x 9.5". $150-250.

Game of Frog by McLoughlin Bros., c. 1880. Framed game board with built-in spinner. Reverse side features Game of Persia. 14.5" x 15". $150-200.

Close-up of mouse serving drinks in the Game of Frog.

Game board to John Gilpin game.

Top right: The Junior Bicycle Game by Parker Brothers, 1897. Bottom of box board, spinner, and paper litho bicycle tokens. 11" x 11.75". *Paul Fink Collection.* $400-600.

Center Right: Game board to The Junior Bicycle Game.

Game of Bicycle Race by McLoughlin Bros., c. 1880. Paper litho over wood with wood tokens and bottom of box board. 10.5" x 19.75". *Paul Fink Collection.* $800-1200.

Paper litho on wood game board for Game of Bicycle Race.

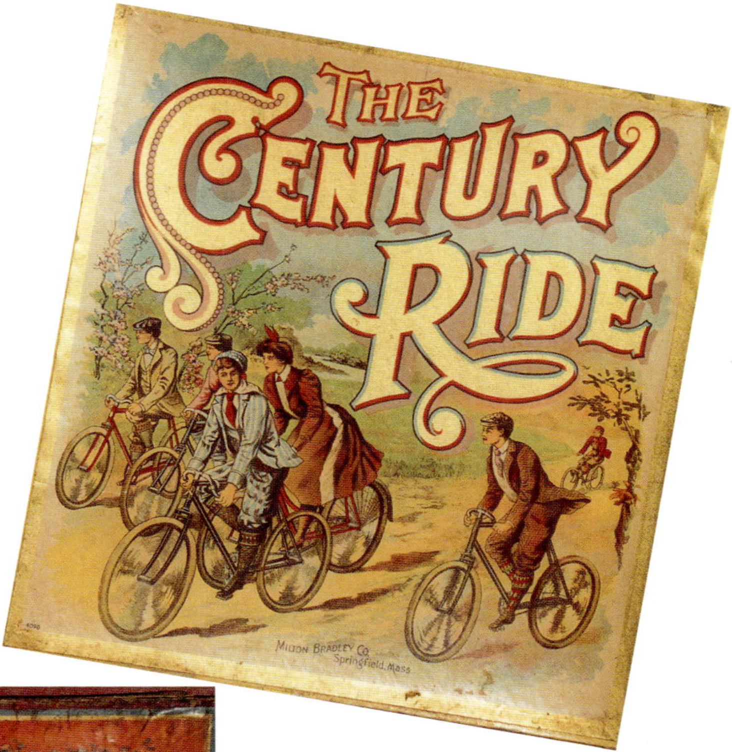

The Century Ride by Milton Bradley, c. 1890. Bottom of box board. 13.25" x 14.25". *Paul Fink Collection.* $200-350.

Top left: Bicycle Game, maker unknown., c. 1890. Bottom of box board, paper spinner, and wood tokens. 7.5" x 7.5". *Paul Fink Collection.* $200-300.

Bicycle Race by McLoughlin Bros., 1895. Paper litho over wood, bottom of box board, paper spinner, and four metal bicycle tokens. 21.25" x 12.75". *Paul Fink Collection.* $400-650.

Paper litho on wood game board to Bicycle Race.

The Street Car Game by Parker Brothers, c. 1890. Bottom of box board, paper spinner, and wood tokens. 15" x 9" *Paul Fink Collection*. $300-500.

Game board to The Street Car Game.

Game of Uncle Sam's Mail by McLoughlin Bros., 1893. Bottom of box map board. 21.5" x 9.5". *Paul Fink Collection*. $300-500.

Game of Traffic by L.H. Nelson Co., Portland, Maine, c. 1900. Bottom of box board. 16.75" x 9.75". *Paul Fink Collection.* $1,000-1,200.

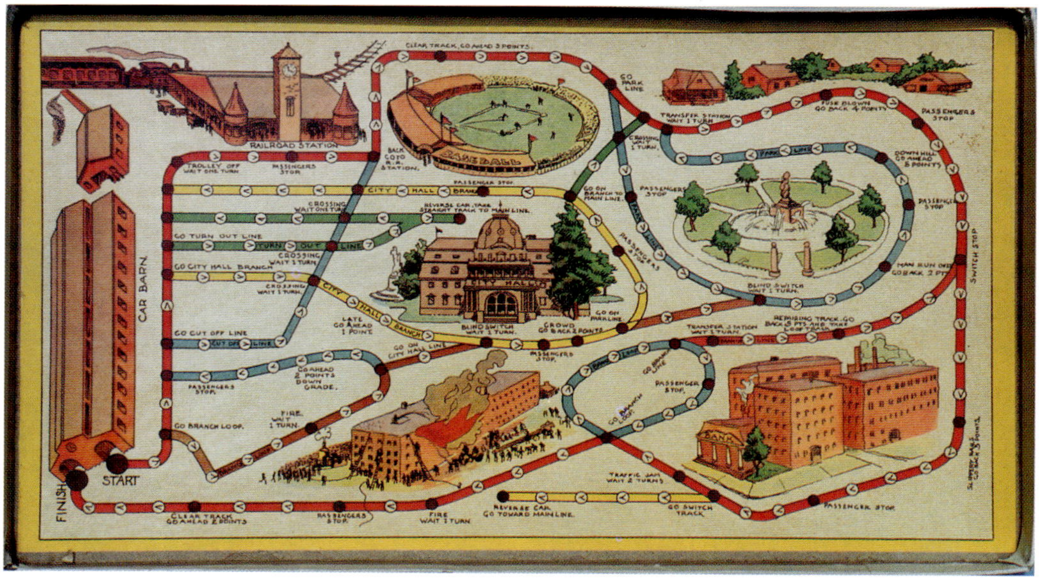

Game board to Game of Traffic.

The Railroad Game by Parker Brothers, c. 1890. Bottom of box track type board, paper spinner, and wood tokens. 20.5" x 10.5". *Paul Fink Collection.* $400-700.

Game of Mail Express or Accommodation by McLoughlin Bros., 1895. Map of U.S. board with paper spinner and metal locomotives. *Paul Fink Collection.* $500-600.

International Automobile Race by Parker Brothers, c. 1905. Bottom of box European map board, cardboard spinner, wood tokens, and small destination cards. 10.5" x 11.5". *Paul Fink Collection.* $1,200-1,500.

Game of The Automobile Race by McLoughlin Bros., 1904. Bottom of box board, cardboard spinner, and wood tokens. 12" x 12". *Paul Fink Collection.* $1,000-1,500.

Game board to Game of The Automobile Race.

Race For The Cup by Milton Bradley, c. 1905. Bottom of box track board, cardboard spinner, and wood tokens. *Paul Fink Collection.* $400-700.

Game board to Race For the Cup.

Merry Christmas Game by Milton Bradley Co., c. 1910. Bottom of box board, cardboard spinner, and wood tokens. 16.5" x 11.5". *Paul Fink Collection.* $500-800.

Game board to Merry Christmas Game.

The Air Ship Game by McLoughlin Bros., 1912. Standard Series #5175. Bottom of box board, cardboard spinner, and wood tokens. 12.5" x 12.5". *Paul Fink Collection.* $600-1,000.

Game board to The Air Ship Game.

Game of To The North Pole by Air-Ship by McLoughlin Bros., c. 1915. Bottom of box track board, cardboard spinner, and wood tokens. 19.75" x 10.5" *Paul Fink Collection.* $500-700.

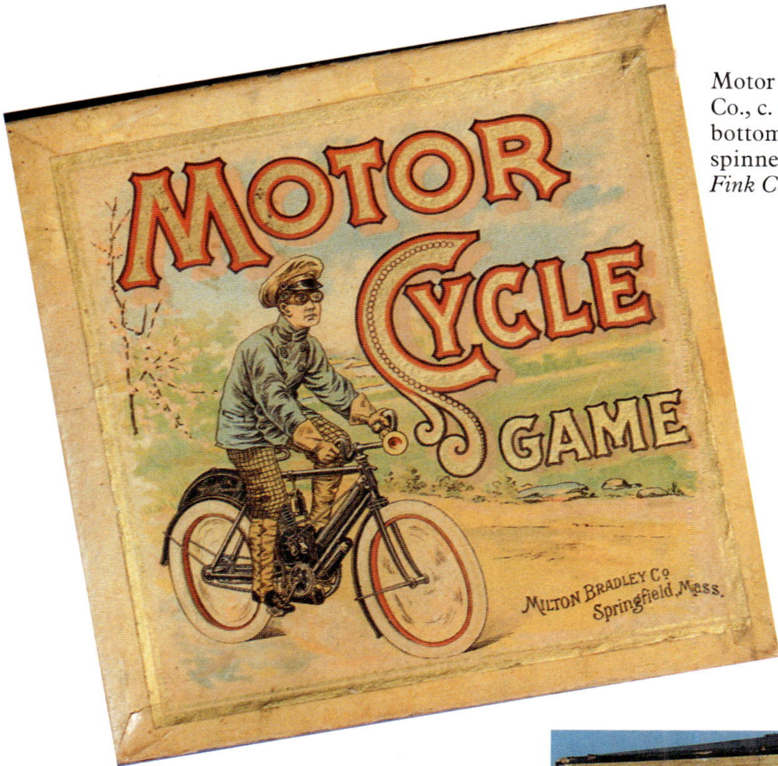

Motor Cycle Game by Milton Bradley Co., c. 1900. Paper litho on wood bottom of box board, cardboard spinner, and wood tokens. 9" x 9". *Paul Fink Collection.* $300-500.

Cross Country Race by Land Air & Sea, manufactured by the Chicago Game Co. (Nut Inc.), c. 1911. Bottom of box board. 12" x 17". *Paul Fink Collection.* $1,500-2,400.

Yacht Race by Milton Bradley, c. 1900. Bottom of box board. 14.25" x 8". $95-165.

International Yacht Race by McLoughlin Bros., c. 1890. Bottom of box board, cardboard spinner, and small wood tokens. 10.5" x 14.5". *Paul Fink Collection.* $250-350.

Game board to The Yacht Game.

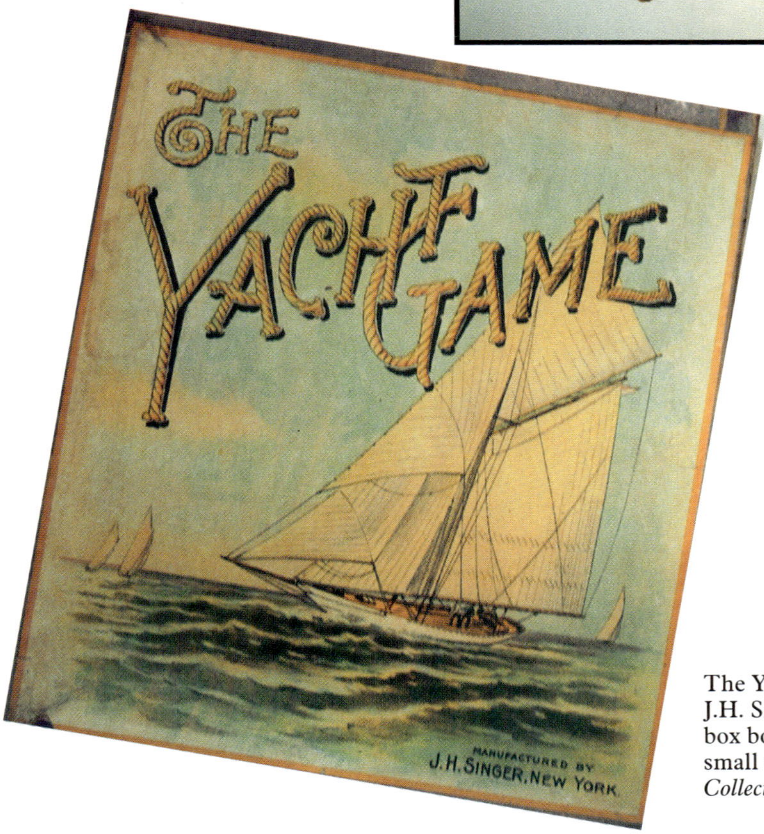

The Yacht Game, manufactured by J.H. Singer, c. 1890. Small bottom of box board, cardboard spinner, and small wood tokens. *Paul Fink Collection.* $250-300.

Hickery Dickery Dock by Parker Brothers, c. 1890. Bottom of box board, cardboard spinner, and wood tokens. 10" x 8". $100-175.

Game board to Hickery Dickery Dock.

Game of District Messenger Boy or Merit Rewarded by McLoughlin Bros., 1886. Paper litho on wood box, separate folding track game board, paper litho on wood spinner, and metal figure tokens. 17.5" x 9". $150-250.

Game of Red Riding Hood by Parker Brothers, 1895. Bottom of box board, cardboard spinner, and wood tokens. 9.5" x 6.5". $100-200.

Game board to Little Boy Blue.

Little Boy Blue by Milton Bradley, c. 1890. #4325. Bottom of box board, cardboard spinner, and wood tokens. 13" x 13". $75-150.

Game of Beauty and the Beast by Milton Bradley, c. 1890. #4004. Bottom of box board, cardboard spinner, and wood tokens. 9" x 9". $75-150.

Game of Leap Frog by Milton Bradley, c. 1900. #4948. Bottom of box board, cardboard spinner, and wood tokens. 8" x 11". $65-135.

Game board to Leap Frog.

Game board to Game of Discovering The North Pole.

Game of Discovering The North Pole. The Merry Christmas Series by McLoughlin Bros., c. 1900. Bottom of box board, cardboard spinner, and wood tokens. 10.5" x 19.5". *Paul Fink Collection.* $1,000-1,500

The Game of Going To The Klondike by McLoughlin Bros., 1898. Bottom of box board. (board and implements are the same as Game of Discovering The North Pole.) 10.5" x 19.5" *Paul Fink Collection.* $1,000-1,500.

Game of the Telegraph Boy by McLoughlin Bros., 1888. Paper litho on wood box, separate track game board, built in spinner, and metal telegraph boy tokens. 19" x 17". *Paul Fink Collection*. $600-800.

Round The World with Nellie Bly. A Novel and Fascinating Game with Plenty of Excitement on Land and Sea by McLoughlin Bros., 1890. Paper litho on wood box, folding game board, spinner, and wood tokens. 8.75" x 16.25". *Paul Fink Collection*. $300-600.

Game board to Round The World with Nellie Bly.

Phoebe Snow by McLoughlin Brothers., c. 1910. Folding game board, cardboard spinner, and wood tokens. 9" x 16.25". *Paul Fink Collection.* $250-350.

Game board to Phoebe Snow.

Whirlpool Game by McLoughlin Bros., c. 1890. #408. Bottom of box board, cardboard spinner, and wood tokens. 7.5" x 7.5". *Paul Fink Collection.* $100-200.

Game board to Whirlpool Game.

The Game of Bombardment by Milton Bradley, c. 1900. Bottom of box board, cardboard spinner, and wood tokens. 11" x 9". $75-125.

Game board to The Game of Bombardment.

Buffalo Hunt by Parker Brothers, c. 1890. Bottom of box board, cardboard spinner, and wood tokens. 5.5" x 5.5". $65-100.

Game board to Buffalo Hunt.

Game of India by Milton Bradley, c. 1900. #4188. Bottom of box board, cardboard spinner, and wood tokens. 9" x 9". $65-100.

New and Improved Fish Pond Game by McLoughlin Bros., 1890. Assorted paper litho fish, wood fishing rods with hooks, and bottom of box fish pond. 18.5" x 9.5". $100-150.

Close-up of bottom of New and Improved Fish Pond Game.

Fish Pond by Clover Manufacturing. Made in Germany, c. 1915. Wood poles with attached magnets, paper litho fish, and folding four-sided fish tank. 12.75" x 10.75". $75-125.

Cardboard fish tank to Fish Pond.

Jeu De La Pêche A Surprises. The Magnetic Fishing Game. Made in Paris, c. 1890. Paper litho on wood box. 16" x 12.5". $200-350.

Implement box for Jeu De La Pêche, paper litho fish, umbrella, frog, and metal fish.

Game playing surface and backdrop for Jeu De La Pêche.

Fish Pond by Milton Bradley, c. 1890. Bottom of box fish pond, wood fishing pole with hook, and paper litho fish. 9" x 12". $75-140.

The Improved Game of Fish Pond by McLoughlin Bros., 1890. Bottom of box fish pond, paper litho fish, and wood fishing rods with hooks. 18.25" x 12.25". $250-350.

Bottom of box to The Improved Game of Fish Pond.

Fish Pond by Clover. Made in Bavaria, c. 1910. Small folding paper litho tank, fishing poles, and paper litho fish. 8" x 7". $50-75.

Fish Pond by Milton Bradley, c. 1900. #4577. Bottom of box fish pond, litho fish, and wood fishing rods with hooks. 18.5" x 9.5". $75-125.

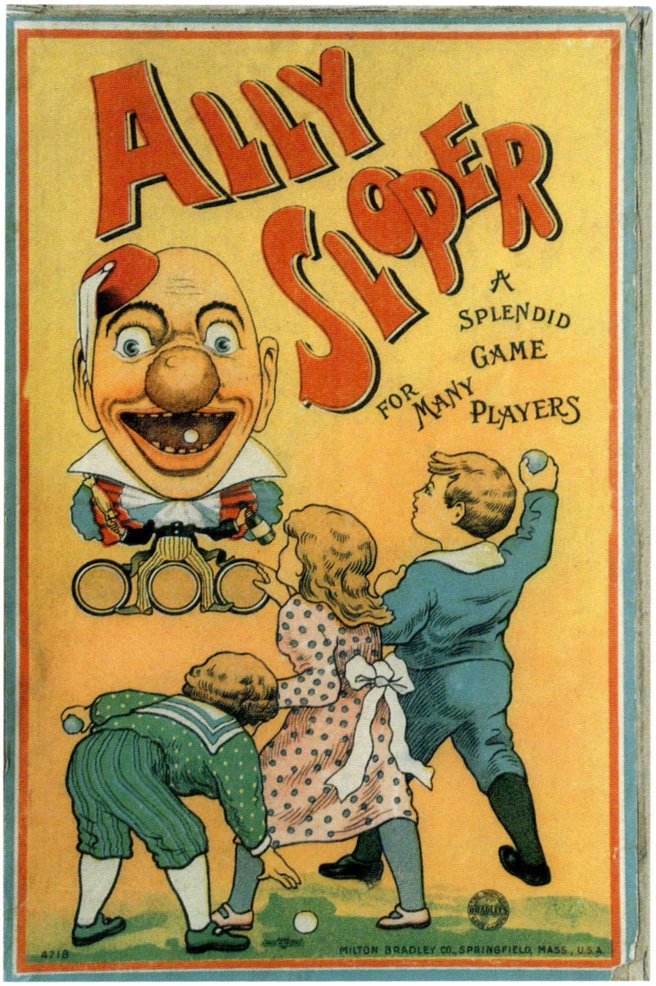

Ally Sloper, A Splendid Game for Many Players by Milton Bradley, c. 1890. #4718. Paper litho marble target game. 7.5" x 12". $100-175.

Target to Ally Sloper.

Swing A Peg by Milton Bradley, c. 1890. #4390. Paper litho on cardboard box with wood ring toss game. Box measures 9.5" x 6.5" . $65-85.

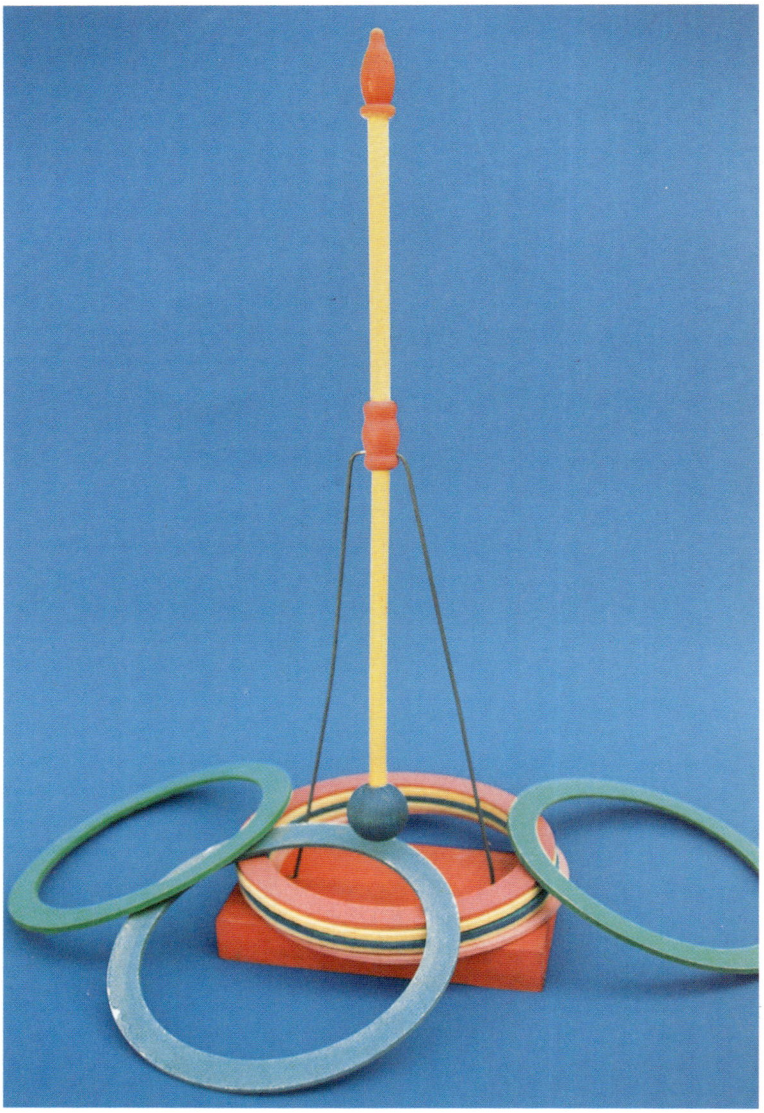

Wooden ring toss to Swing A Peg.

Game of Jitney by Milton Bradley, c. 1920. #4762. Bottom of box marble game. 12.25" x 12.25". *Paul Fink Collection.* $250-500.

Paper litho on wood bottom of Game of Jitney.

Spear's Blow Football Improved Edition. Copyright in the United Kingdom by J.W. Spears & Sons. Manufactured at the Spear Works Bavaria, c. 1900. Six metal pipes, small ball, and two paper litho goal tenders. Box measures 10.5" x 6". $75-150.

Sailboat to Mechanical Yacht Race.

Quoits by Milton Bradley, c. 1890. #470. Ring toss game with wood base and rope rings. 13.75" x 7.25". $50-75.

Sailboat to Mechanical Yacht Race.

Mechanical Yacht Race, The Latest Game. British manufacture, c. 1910. Tabletop sailboat race game. Game includes two American and two British paper litho sailboats, and four pieces of string with hooks attached at each end. Box measures 11.5" x 8.5". $100-150.

Fine Table Croquet by R. Bliss, c. 1890. Paper litho on wood box, eight small mallets, wickets, eight wooden balls, and two posts. 17" x 6". $170-250.

Inside of box to Fine Table Croquet.

Familiar Objects, A Game For Children by Spears, c. 1890. Lotto type game with color litho boards and matching tiles. 11" x 8.5". $100-150.

Tiles to Familiar Objects, A Game For Children.

Lotto by McLoughlin Bros., c. 1890. Paper litho on wood box, glass number covers and color lithograph playing boards. 5.5" x 8". *Courtesy of Austern's Antiques*. $60-100.

Jack Straws by Milton Bradley Co., c. 1890. Multi-colored wood straws with wood handled hook. 6" x 5". $40-50.

Cards to Lotto.

Jack Straws New Edition by Parker Brothers, c. 1890. #163. Multi-colored wood straws and wood handled hook. 6" x 5". $40-50.

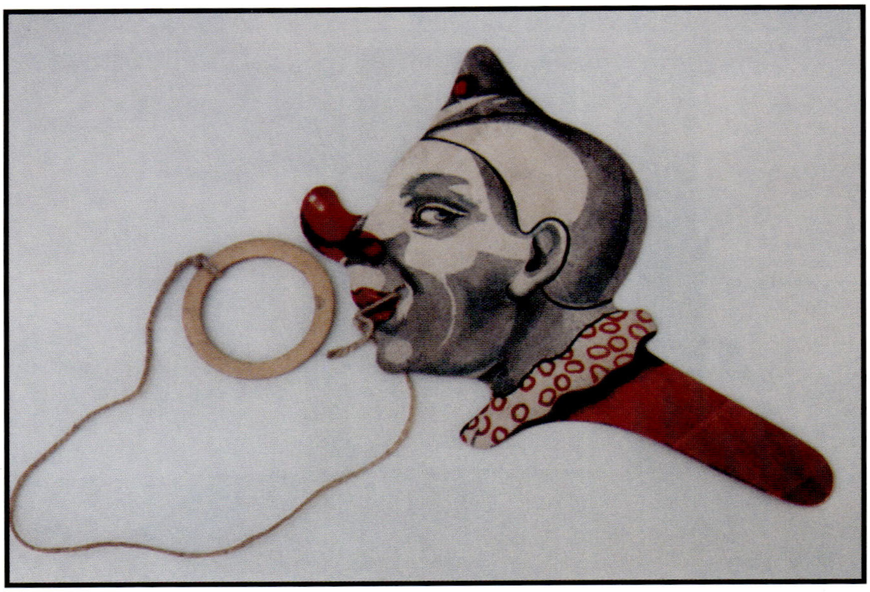

Clown Ring Toss Game, c. 1900. Paper litho advertising premium. $20-30.

Tiddledy Winks. Cup and celluloid disks. 5.25" x 5.25". $40-50.

Magic Dots For Little Tots by Milton Bradley, 1907. #4482. Set includes black and white playing boards, plus colored dots that you fill in to make picture. 4.5" x 6.5". $40-50.

Little Folks Color Kit by Milton Bradley, c. 1890. #4396. Contains watercolor paints, paintbrush, and painting cards. 11" x 14". $60-120.

Dandy Paint Box by Milton Bradley, c. 1890. #4537. Flip top box with watercolor paints, brushes, and painting sheets. 9" x 13". $50-75.

The Little Postcard Painters, unmarked, c. 1900. Watercolor paints, blank coloring cards, and color lithograph pictures to copy. 12" x 11" $40-75.

Inside of box to The Little Postcard Painters.

Geographical Drawing Teacher by Parker Brothers, c. 1900. Map drawing game. 7.5" x 10.5". $40-50.

Paper litho sliding top wood paint box, c. 1900. 4.25" x 5.75". $50-75.

Our Pets Painting Outfit by Milton Bradley, c. 1895. Embossed painting cards, stencils, brush, and paints. 13" x 8.25" $60-100.

Paper litho on wood animal stamps, c. 1900. $20-40.

Paper litho on wood Child's Pencil Box. Made in Germany, c. 1905. 8.5" x 2.5". $30-60.

Paper litho on wood Child's Pencil Box. Made in Germany, c. 1900. 8.5" x 2.5". $30-60.

Little Red Riding Hood Went Through the Wood Pencil Box, c. 1890. 8.5" x 2.5". $30-60.

Paper litho on wood pencil box. Made in Germany, c. 1900. 8.5" x 2.5". $30-60.

Bead Stringing by Milton Bradley Co., c. 1910. 6.5" x 6.25". $20-40.

Paper litho on wood paint box, c. 1890. 7.5" x 4.5". $55-65.

The Donkey Party by J.H. Singer, c. 1890. Combination game and pin the tail on the donkey. Bottom of box board, spinner, wood tokens, printed fabric donkey, and fabric donkey tails. Box measures 8.5" x 7.25. $50-100.

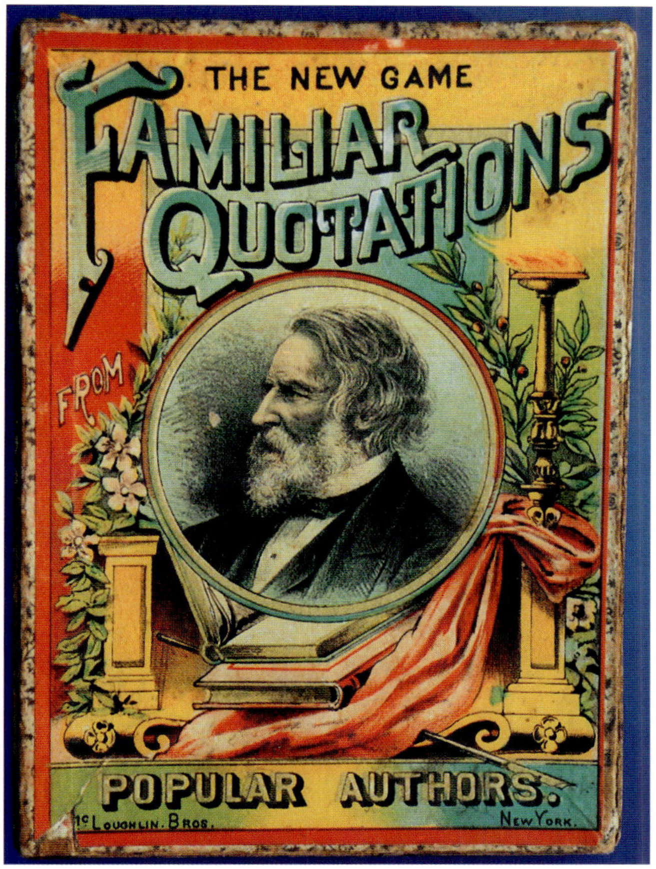

The New Game Familiar Quotations, Popular Authors by McLoughlin Bros., c. 1890. Black and white reading cards. 4.5" x 6". $25-40.

'Round The World Joe by Parker Brothers, c. 1890. A reading game. 3.5" x 5.0". $40-60.

Home History Game For Boys and Girls by Milton Bradley, c. 1890. #4273. A reading game. 5"x 6.5". $30-50.

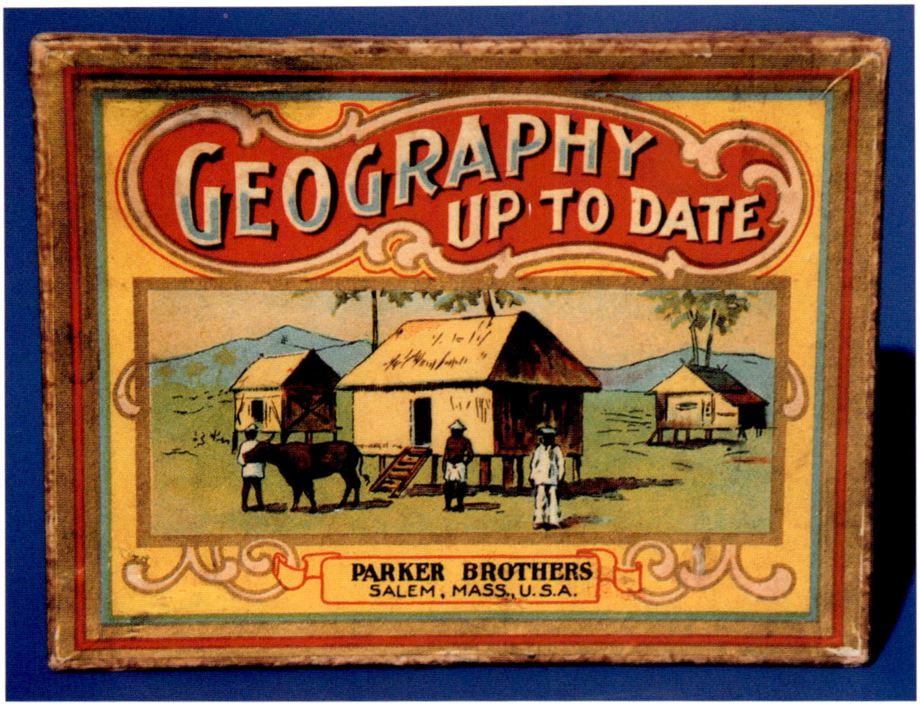

Geography Up to Date by Parker Brothers, c. 1900. World Trivia card game. 6.75" x 5.25". $30-50.

The Comical Game of Who by Parker Brothers, c. 1900. A reading game. 5" x 6.5". $25-40.

How Silas Popped the Question. Unmarked, c. 1890. A reading game. 5" x 6". *Courtesy of Austern's Antiques*. $30-50.

Peter Coddle and His First Trip to New York by McLoughlin Bros., c. 1890. A narrative reading game where each player fills in the blanks depending on what words they pick. 3.5" x 4.5". $45-65.

Peter Coddle's Trip by Milton Bradley, c. 1900. #4380. A reading game. 6" x 8". $30-50.

My Mother Sent Me to the Grocery Store by Parker Brothers, c. 1900. #364. 5.25" x 7". $45-65.

The Fortune Teller Game by Milton Bradley, c. 1900. Box cover is same as spinner. Includes booklet and spinner. 7.5" x 8.5". $50-75.

"Chantecler" the Great Card Game by Parker Brothers, c. 1910. Paper litho box and cards. 5.5" x 4". $40-55.

Cards to Old Maid.

Old Maid by Milton Bradley, c. 1910. Paper litho box and cards. 5.5" x 6.75". $40-65.

The Merry Game of Old Maid by McLoughlin Bros., 1892. Elite Edition. Paper litho picture cards. 5.5" x 6.5". *Paul Fink Collection.* $75-110.

Old Maid by Milton Bradley, c. 1900. #4114. Picture cards consist of nine sets of two and an Old Maid card. Box measures 5.5" x 6.5". $50-75.

The Jolly Old Maid by Parker Brothers, c. 1890. #371. Paper litho cards. Box measures 5" x 6". $35-55.

The Popular Game of Old Maid by United Game Co., c. 1900. #174. Box measures 6.5" x 8.5". $50-100.

Cards to The Popular Game of Old Maid.

Three Bears by Milton Bradley, c. 1900. #4770. Two-color picture cards. 6" x 4.5". $75-125.

Game of Robinson Crusoe by Parker Brothers, 1895. Set includes two of each card pictured. Box measures 4" x 5.5". $50-75.

Cards to Game of Robinson Crusoe.

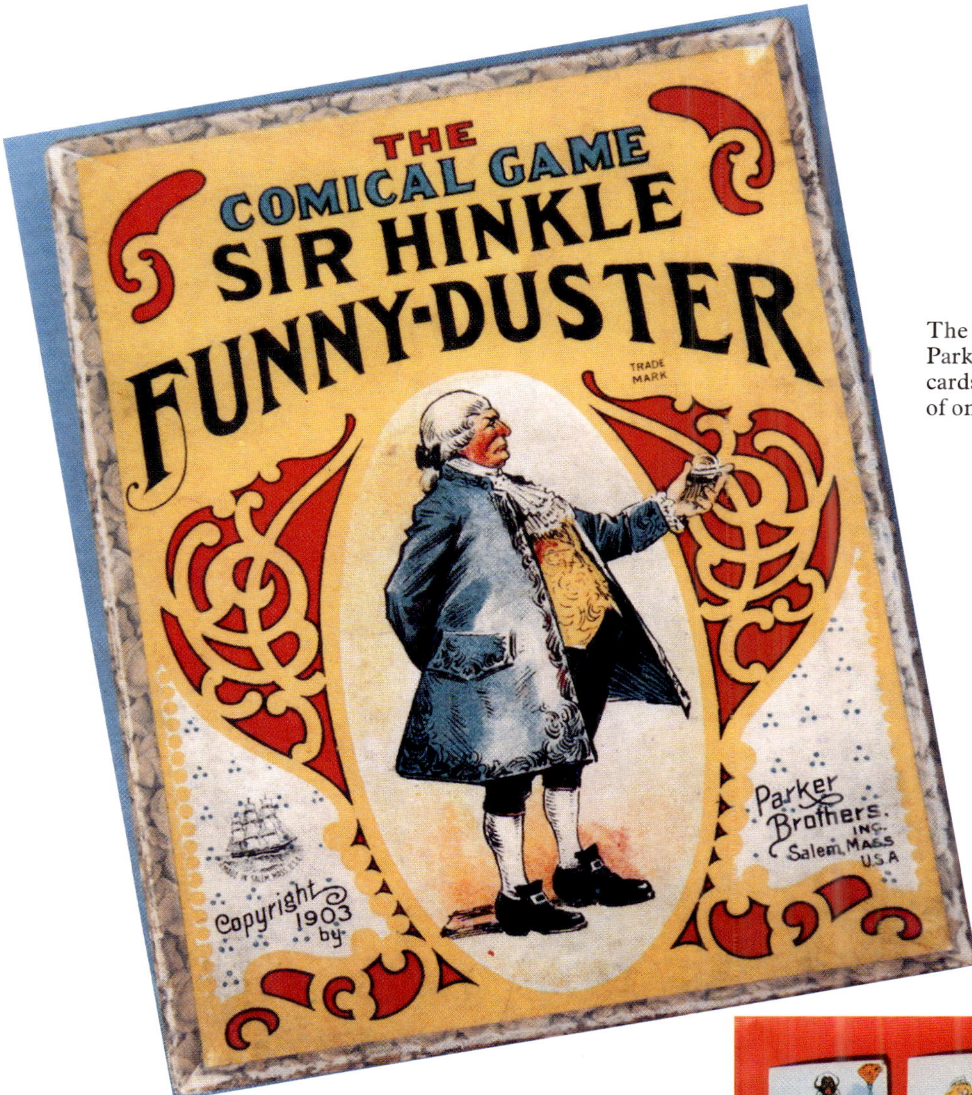

The Comical Game of Sir Hinkle Funny-Duster by Parker Brothers, 1903. Different sets of "Family" cards. The object of the game is to get a complete set of one family. Box measures 7" x 5.25. $55-75.

Cards to The Comical Game of Sir Hinkle Funny-Duster.

The Premium Game of Logomachy or War of Words by McLoughlin Bros., 1889. Set of seventy-two color picture cards with letters. Wood box. 5" x 6". $50-75.

Cards to The Premium Game of Logomachy or War of Words.

The Jolly Game of Snap by Parker Brothers, c. 1900. Brownie type picture cards. Box measures 6.5" x 8.5". $50-75.

Cards to The Jolly Game of Snap.

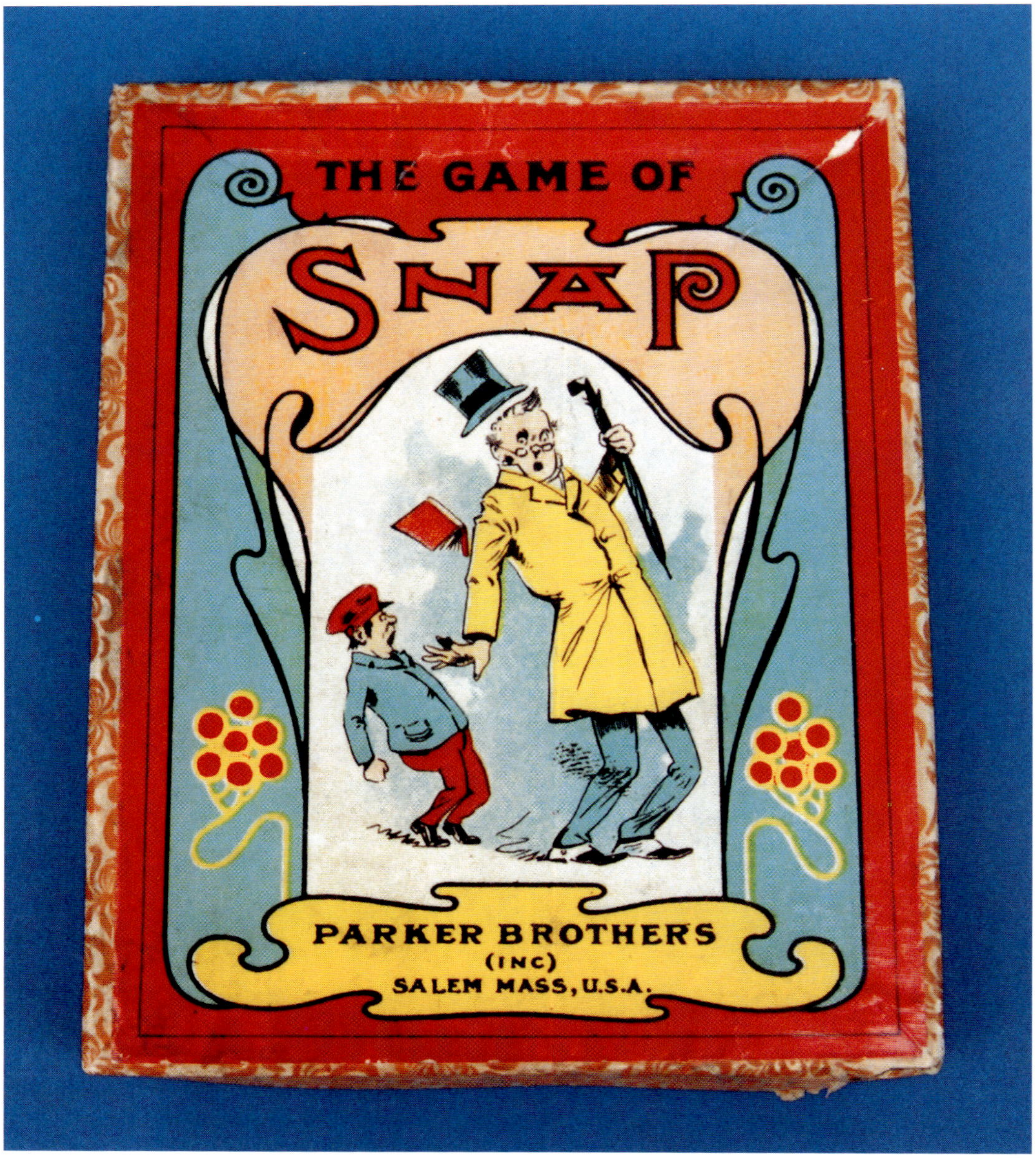

The Game of Snap by Parker Brothers, c. 1900. Nine sets of three picture cards each. A matching game. Box measures 5" x 6". $40-60.

Cards to The Game of Snap.

Bicycle Cards, A Game by Parker Brothers, 1897. Picture card game. 6.5" x 5". *Paul Fink Collection.* $150-250.

Cards to Bicycle Cards A Game.

Game of Shoot the Hat by McLoughlin Bros., 1892. Punch and Judy Series. Picture card game. Box measures 4" x 3". *Courtesy of Austern's Antiques*. $75-150.

Cards to Game of Shoot the Hat.

Jungle Jinks, Jumbo and the Bull by Thomas De La Rue & Co. London, c. 1890. Picture cards illustrated by Constance Standard Chapman. 48 cards. 2.5" x 4". $75-125.

Cards to Jungle Jinks.

The Black Cat Fortune Telling Game by Parker Brothers, 1897. Picture card game. 6.5" x 5". $75-150.

Cards to The Black Cat Fortune Telling Game.

Card Houses by J.W. Spear & Sons, London, c. 1900. Printed in Bavaria. Picture card building game. 5.5" x 4.5". $40-60.

Cards to Card Houses.

Dr. Busby by Parker Brothers, 1910.
Picture card game. 7.5' x 5.5". $45-75.

Game of Dr. Busby by Milton Bradley, c. 1890.
#4039. Paper litho picture cards. 5" x 6". $45-75.

Cards to Dr. Busby.

Game of Nations by Milton Bradley, c. 1890. Paper litho on cardboard picture cards of different countries. 5" x 6". $25-50.

Cards to Game of Nations.

The Game of Nosey by McLoughlin Bros., c. 1900. Paper litho on wood box, papier mâché nose with attached wire glasses, and colored playing cards. 8.5' x 4.75". *Paul Fink Collection.* $300-600.

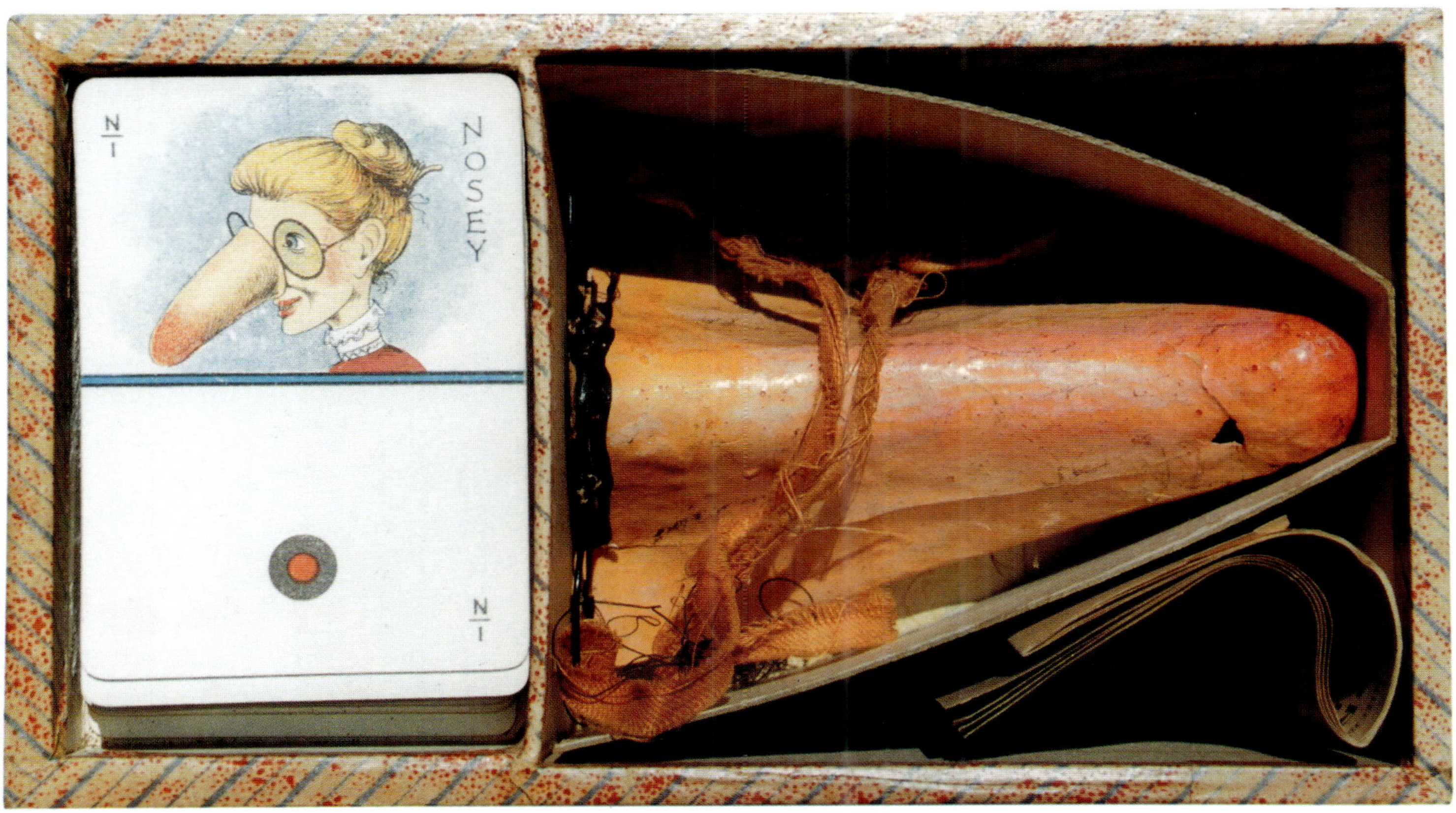

Cards and Nose to The Game of Nosey.

Chapter Two

Toys

Toys at the end of the nineteenth century were usually constructed of wood covered with graphic, brightly colored lithographed paper. Toys consisted of wooden boats, nine pins (in Europe) and ten pins (in America), push and pull toys, clockwork and mechanical toys, transportation vehicles, educational toys, miniature theaters, painting sets, panoramas, and furniture for dolls.

The foremost names in children's paper litho toys were W.S. Reed, Charles Crandall, The McLoughlin Brothers, The Gibbs Manufacturing Company, R.W. Bliss, Milton Bradley, and Parker Brothers. In an 1881 advertisement for *Wide Awake* magazine, Crandall's described their Treasure Box: "This set is one of Crandall's latest and best inventions. It is not a single toy, but a dozen in one, comprising a Wagon, Top, Bedstead, 2 Chairs, Wheelbarrow, Mallet, Bench, Table, Puzzle, Set of A B C Blocks, a group of nine Domestic Animals and a pretty Suburban Village. A Most delightful combination for the little folks. Price $1.00 Too large to go by mail expressive to be paid by receiver."

In addition to the American manufacturers, Germany exported many toys and was known for its excellence in using clockwork mechanisms combined with wonderful lithography. England and France were also contributors to the paper litho toy market at the end of the nineteenth century.

Left and opposite page: American and European Menagerie by Bliss Mfg., c. 1895. Paper litho on wood. This set was offered in three sizes. The small set did not have a rider. 7.5" tall. (Set too rare to determine a price.)

AMERICAN · AND · EUROPEAN · MENAGERIE

AMERICAN · AND · EUROPEAN · MENAGERIE

Christmas Moving Picture by McLoughlin Bros., c. 1890. Shadow box displaying a paper lithograph diorama of Christmas day. Five moving figures. 13.5" x 17.5" x 5". $7,000-9,000.

London General Omnibus Co. Limited, c. 1890. Paper litho on wood horse drawn double decker carriage advertising Sapolio cleaner. (Too rare to determine a price.)

Horse Drawn Fire Reel by Bliss, c. 1890. Paper litho on wood wagon, fire wheel, and two figures. Pulled by a single horse. 19.5 x 8". $4,000-5,000.

Pull For the Shore by W.S. Reed, c. 1881. When pulled forward the man on the stern rings the bell as the man in the front plays the cymbals (figure in bow is missing.) The two men in the middle move to the music. $6,000-9,000.

Close up of scene on side of Pull For the Shore.

Bradley's Menagerie and Aquarium by Milton Bradley, c. 1880. Paper litho on wood panorama scroll toy. One side shows scenes from the menagerie, the other for the aquarium. (Too rare to determine a value.)

View of Parrots in Bradley' s Menagerie and Aquarium.

Horse Drawn Fire Pumper by Reed, c. 1890. Paper litho on wood. $3,500-5,000.

Santa Sleigh by Bliss, c. 1900. Paper litho on wood. 18" x 8". $3,700-4,600.

The Sealark by Bliss, c. 1895. Paper litho on wood and cardboard tug boat, also came with a barge. Wheels on bottom of ship. 19.5" x 16". (Too rare to determine a value.)

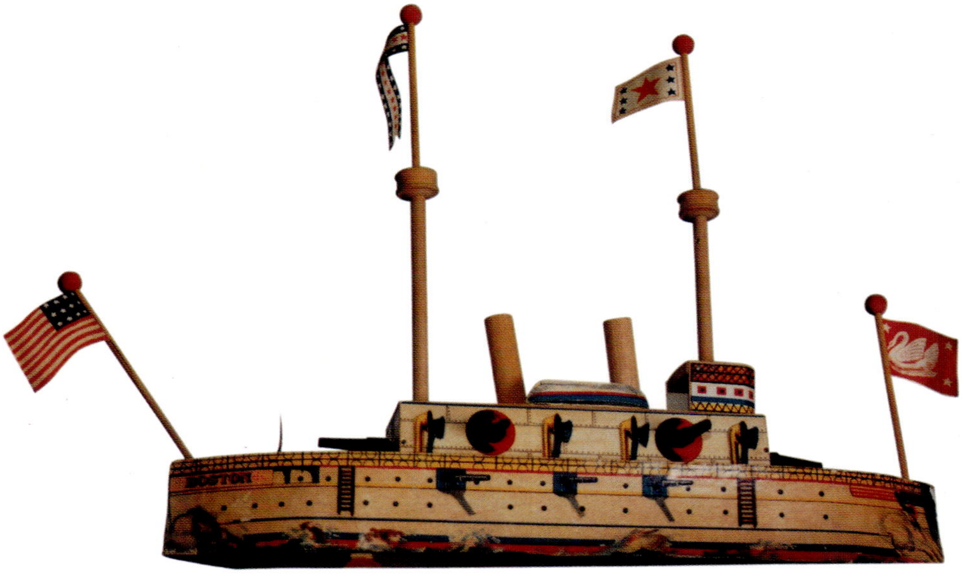

The Boston. Unmarked, possibly by Bliss, c. 1900. Paper litho on wood and cardboard ship with four wooden canons, two crow' s nests, two masts, and two flags. 19.5" x 16". $2,000-2,875.

The Boston sailboat. Unmarked, possibly Bliss or Reed, c. 1900. Paper litho on wood ship. 17" x 16". $250-500.

Paper litho on wood Ark, made in Germany, c. 1890. Painted roof is hinged, animals stored inside. 22" x 6". (Too rare to determine a value.)

Paul's Soap Circus, c. 1900. Paper litho on wood ark with removable top. Housed large animal shaped soaps. 8" x 6". $200-300.

Back view of Paul's Soap Circus.

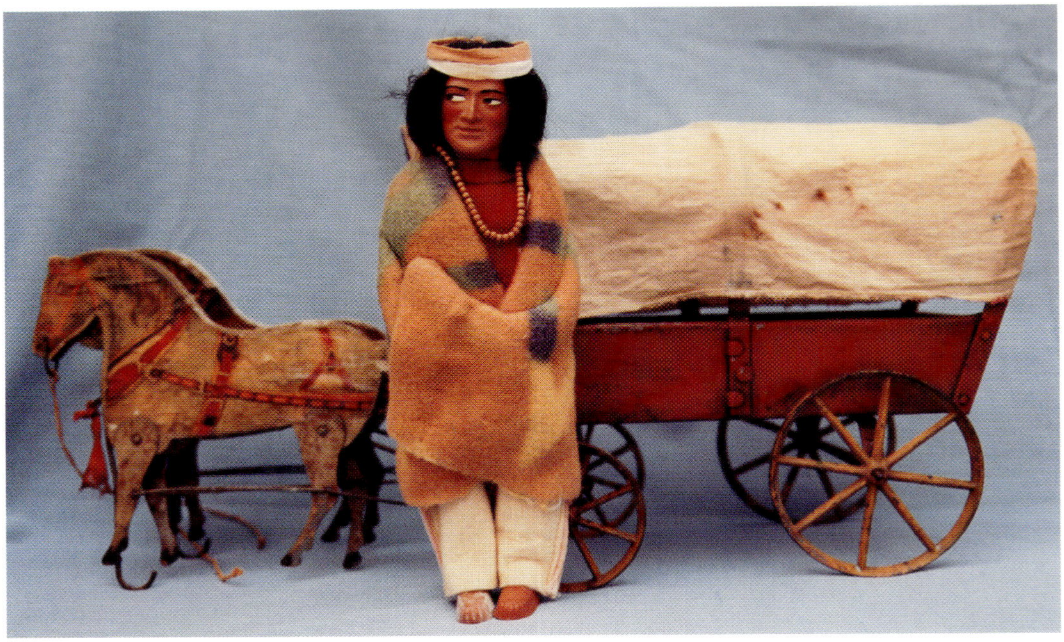

Horse drawn covered Gypsy Wagon by Gibbs, c. 1900. Red tin wagon with four spoked wheels pulled by two paper litho on wood horses. The horses have articulated legs. 21" long. $250-325.

Broadway Cable Line, c. 1900. Paper litho on wood mechanical cable car that moves backwards and forwards. Sign on top switches four ways. 18" x 8.5". (Too rare to determine a value.)

Close up of passengers in Broadway Cable Line.

Paper litho on wood bagatelle game by J.H. Singer. March 25th, 1890. 13.5" x 6.5". $200-350.

Bagatelle Game, c. 1890. English. Paper litho on wood. 13" x 6.5". $150-300.

Punch and Judy Theater by Spear's, c. 1890. Stand-up theater with paper litho figures on wire. Stage sets up on bottom of box. 8.5" x 12". $250-350.

Theater to Punch and Judy Theater.

Punch and Judy Blow Toy by Ives, c. 1890. Blowing into pipe causes figures to move. 4.75" x 3.75". (Too rare to determine a value)

Christmas Mechanical Toy, c. 1890. Made in Germany. Mechanical clockwork chromolithograh die-cut Santa pulls clown our of sack, clown' s arms and legs move, Santa moves his head and child's head moves. 15" x 17". $6,000-8,000.

The Chicken Dispute, c. 1890. Unmarked. Paper litho on wood action toy. Pushing wood levers on back causes the chickens to peck. 7.5" x 4". $200-300.

Chicken Pip Squeak, c. 1900. Paper litho on wood roof. 4.5" tall. $45-95.

Ringo by Milton Bradley, c. 1910. Chicken Ring Toss by Milton Bradley, c. 1900. Object of the game is to ring the tails of the chickens and rooster. $45-65.

Articulated Alligator by Wilder Mfg., c. 1910. 17" x 4.4". $30-50.

The House That Jack Built Bank, c. 1890. Paper litho with shell roof. 6" x 5". $500-700.

Paper litho Bank, c. 1890. Paper litho on wood box. Box opens with key and has slot on top. 3.25" x 4.5". $50-75.

Musical Rattle, c. 1900. Paper litho on wood musical rattle. 8". $50-100.

Ornate Doll's Rocking Chair by Reed, c. 1890. Paper litho on wood. 6" x 11" x 6". (Too rare to determine a value.)

Block Rattle, c. 1900. Paper litho on wood musical rattle. Block shape features a clown on each of four sides. 2" x 5.5". $40-60.

The Alphabet Chair by Forbes of Boston, c. 1890. Paper litho on wood doll's high chair. 5.25" x 12". $600-800.

Alphabet Rocking Chair by Reed, c. 1890. Doll's paper litho on wood chair shows two children and a dog having a tea party. 7" x 17". $600-1200.

Seat to Alphabet Rocking Chair.

The Spelling Chair by Reed, c. 1890. Paper litho on wood rocking chair. 6" x 14.5". $600-800.

Alphabet High Chair by Reed Toy Co., c. 1890. Paper litho on wood. 6.5" x 23". $350-600.

Doll Trunk, c. 1890. Paper litho on wood. 12" x 7.5". $250-400.

Bear Musical Push Toy, c. 1910. Paper litho on wood with cast iron wheels. Has wooden horse head at end of wood stick. 8" x 35". $400-600.

Close up of wooden horse head to Bear Musical Push Toy.

Butterfly Push Toy by Gibbs, c. 1890. Paper litho on tin wings, painted body, cast iron wheels, and wood handle. Wings flap with pushed. 9" wing span. $150-200.

Large Child's Rocking Horse, c. 1890. Paper litho on wood. Very rare size, possibly French. 43" x 24" high. (Too rare to determine a value.)

Child's Piano by F.D.C., Glasgow, c. 1900. Paper litho on wood working piano . Litho on all sides. 10" x 6". $250-350.

Side view of Child's Piano.

Child's Piano by Bliss, c. 1890. Paper litho on wood child's working piano. *Judy McClymond Collection*. $150-300.

Boys Union Tool Chest by Bliss. c. 1900. 12" x 6". $60-140.

Bliss Union Tool Chests for Boys, c. 1900. Includes original wood and metal tools. 12" x 6". $45-75

Elite Tool Chests For Boys. Manufactured by American Manufacturing Concern, c. 1900. No. 425. Paper litho on wood. Dovetailed wooden sides. 12" x 6". $40-60.

Tin Musical Crank Toy, c. 1900. Paper litho on tin. 2.5" x 2.5". $100-150.

Cat and Mouse Wood Music Box, c. 1890. Paper litho on wood musical crank toy. 3" x 2.25". $75-150.

Close up of Child's Xylophone, c. 1900. Paper litho on wood. 7.75 x 1.75". *Cathy Cook Collection*. $55-75.

Paper Lithographed Accordion. Made in Germany. c. 1890. Paper on cardboard. 3" x 4.5". $30-50.

Jack and Jill Ten Pin Blocks by Bliss Manufacturing Co., 1893. Paper litho on wood. Box measures 6.75" x 7.75". Each pin measures 1.25" x 7". $1,500-2,000.

Contents to Jack and Jill Ten Pin Blocks.

Tiny Tot's Nursery Skittles. British made, c. 1910. Paper litho on wood. Roll wood balls to knock down pins. Nine Pins (Europeans used nine pins while Americans used ten pins.) Box measures 11.5" x 9.5". (Too rare to determine a value.)

Contents to Tiny Tot's Nursery Skittles.

Military Ten Pin Soldiers by Ives, c. 1890. Paper litho and paint on lathe turned wood. Five red and five blue soldiers with wood balls. $3,000-4,500.

Darkey Five Pins by Milton Bradley, c. 1900. #4595. Paper litho on cardboard. Box measures 6" x 8.5". Pins are 6.5" high. $400-500.

Five pins to Darkey Five Pins.

Soldier Ten Pins by McLoughlin Bros., c. 1890. 9" x 7.5". $75-150.

Contents to Soldier Ten Pins.

Jolly Darkie Ten Pins, by McLoughlin Bros., c. 1890. Paper litho on cardboard bowling type game. 9.5" x 15". Each pin measures 8" high. (Too rare to determine a value.)

Pins to Jolly Darkie Ten Pins.

Comical Character Nine Pins, c. 1890. European paper litho on wood skittles game. Each figure measures 7" tall. (Too rare to determine a value.)

Top center:
Elephant roller skating from Animal Ten Pins.

Top right:
Frog with walking stick from Animal Ten Pins.

Animal Ten Pins by Milton Bradley, c. 1890. Large paper litho on cardboard animal figures with wood bases. 3.5" x 9.5". Boxed set $400-600. $20-30 each. Shown here is Cat playing tuba from Animal Ten Pins.

Bottom center:
Bull shoe shine from Animal Ten Pins.

Bottom right:
Bear playing violin from Animal Ten Pins.

Aunt Sally by Parker Brothers, c. 1880. Ring toss game. Paper litho on wood face. Box depicts Aunt Sally in black and white. Box measures 6.5 x 16". Aunt Sally measures 4" x 15". $150-300.

Close up of face to Aunt Sally.

Trap Shooting by Milton Bradley, c. 1900. Paper litho on cardboard targets, tin spring gun, and wood bullets. 8" x 9". $100-150.

Shooting target to Trap Shooting.

Mother Goose Target Game by Reed, c. 1890. Paper litho on wood. Roll wooden balls to hit the knobs and then the targets pop up. 18" x 11". $500-700.

Dancing Darkies, c. 1890. Paper litho on wood jigger toy. 4" x 8". (Too rare to determine a value)

Old Kentucky Home by Reed, c. 1890. Paper litho on wood articulated figures. Crank activates six dancers. $4,500-6,500.

Chapter Three

Blocks and Puzzles

One of the most artistically appealing categories in paper litho toys is the category of blocks. Whether alphabet blocks, nesting blocks, or cube blocks that form different pictures, paper lithographed blocks are one of the easiest to display. The McLoughlin Brothers, Charles Crandall, W. & S.B. Ives, Milton Bradley, Bliss, and various German manufacturers produced many of the blocks in the late 1880s. An 1889 McLoughlin Brothers Toy Catalog describes a set of stacking blocks as follows: "A set of twenty-four flat blocks; size, 3 1/4 inches square and 1 inch thick. Elegantly printed in colors and gold. They contain an alphabet and object lessons on one side, and pictures of animals, birds, etc., on the other. On the ends are simple words, beginning with the letter of the alphabet on the block. Put up in a handsome box, with very showy and attractive illuminated label. $1.50."

Milton Bradley also described their blocks in a toy catalog from 1882: "Twelve large cube blocks in red, white and blue oil colors, have on 48 sides 36 illustrations of familiar rhymes from Mother Goose, and all the alphabet and numerals and on the other sides various colors and shades, including the primary and secondary colors and representative tints. The illustrations are all from new and original designs, and the color sides form a valuable object lesson in colors. Each set is accompanied in the box by a book containing the Rhymes which are illustrated on the blocks, so that the whole makes a most fascinating combination of colors, blocks and

ABC Blocks by Whitney Reed Chair Co., c. 1890. Paper litho on wood set of die-cut alphabet blocks. Box measures 10" x 13". $2,000-2,500.

jingles. In a polished wooden box with beautiful chromelithographic label." Finding a complete set of alphabet blocks or stacking blocks is a wonderful find. Just displaying a few paper litho blocks on a shelf, however, can also be a visually rewarding experience.

Dissected puzzles of the late nineteenth century were simply paper lithographed pictures applied to a heavy paper board and then cut into slices. In some cases a dissected puzzle could mean a sliced or jigsaw puzzle. The jigsaw puzzle was invented in the 1870s when the jigsaw (a machine with which an irregular pattern could be cut) was invented. Early jigsaw puzzles used wood with chromolithographed paper applied on top. Puzzles were a category that both children and adults enjoyed. Many of the early sliced puzzles were educational in nature; some taught spelling using fairy tales and animal names, for example.

In the 1889 McLoughlin toy catalog sliced puzzles were called "puzzle strips:" "Each box contains six picture puzzles, 8 1/2 x 11 inches, cut up in strips. Both sides of the strips can be used in forming the puzzles. The pictures consist of fine delineations of the animal world. The coloring is true to nature and the pictures are bright and showy. Put up in a fine wood box with varnished label. Retail 30 cents each." Milton Bradley also described dissected puzzles in their 1882 toy catalog: "Dissected puzzles are always attractive to children, and the effort necessary to put them together develops their constructive genius. When the design is a map the element of geography is embraced, and when the picture represents some mechanical construction with the names of the parts, a new branch of information is added."

Christmas Alphabet Blocks by McLoughlin Bros., c. 1890. Paper litho on wood. Twenty-four block set. 13" x 13". $1,000-1,200 with box. $400-500 without box.

Alphabet Blocks by McLoughlin Bros., c. 1890. These are the reverse side of the Christmas scene.

Letter "U" to Alphabet Blocks.

Letter "N" to Alphabet Blocks.

Read and Learn Blocks by Bliss, c. 1890. Paper litho on wood cylinder alphabet blocks. They came in either six, twelve or twenty-four block sets. Each block measures 1.75" x 4". (Too rare to determine a value.)

Back view to Read and Learn Blocks.

Santa Sleigh with Blocks, c. 1890. Paper litho on wood sleigh with small alphabet blocks. 13" x 5.25". (Too rare to determine a value.)

Gold Lettered Alphabet Blocks, c. 1890. Paper litho on wood. 2" x 2". $135-165.

Nursery Joy For Girls & Boys by Bliss Manufacturing Co., c. 1885. Paper litho on wood alphabet blocks. Box opens to a checkerboard table. Bliss made these in two sizes. Each block measures 2" x 1.25". $500-800.

Blocks from inside box to Nursery Joy For Girls & Boys. Reverse side of blocks is a set of dominoes.

Peepshow ABC Blocks by McLoughlin Bros., 1884. Paper litho on wood four-sided blocks with embossed pictures under glass. Box is 10.25" x 8". Each block measures 1.4" x 1.4". (Too rare to determine a value.)

Blocks to Peepshow ABC Blocks.

Our Pets Picture Book Blocks by R. Bliss Mfg. Co., 1891. Small paper litho on wood alphabet blocks shaped like miniature books. Box measures 5.5" x 12.5". Each block measures 2.5" x 3". $1,200-1,600.

Close up of letter "A" block to Our Pets Picture Book Blocks.

Side view of alphabet blocks.

Punch and Judy Alphabet blocks, c. 1890. Small blocks with train pictures on the reverse side. Most likely made by McLoughlin Bros.(Too rare to determine a value.)

Little Sweetheart Embossed Blocks by Bliss, c. 1895. Die-cut embossed paper litho by Raphael Tuck on wood blocks. These came in assorted sets of six die-cuts per story. The boxed set pictured has twelve blocks, four stories of six scenes each: Jack and the Beanstalk, The 10 Little Niggers, Water Circus, and The Circus. (Too rare to determine a value.)

Set of blocks to Little Sweetheart Embossed Blocks.

Drum Picture Blocks by Bliss, c. 1896. Paper litho and cardboard on wood, drum shaped blocks. 2.75" in diameter. $1,500-2,200.

Close up of Drum block.

Fire Department & ABC Picture Blocks by McLoughlin Bros., 1893. Paper litho on wood blocks. Box measures 8.5" x 13". Each block measures 2.25" x 4". $1,600-$2,400.

Blocks to Fire Department & ABC Picture Blocks.

Shadow ABC Blocks by R. Bliss Mfg. Co., 1890. Paper litho on wood blocks. Box measures 8" x 11". Each block measures 1" x 1.5". (Too rare to determine a value.)

Blocks to Shadow ABC Blocks.

Playmates ABC Blocks. Made in Germany, c. 1890. Paper litho on wood blocks. Box measures 5" x 6.75". Each block measures 1.5" x 1.5". $85-150.

Blocks to Playmates ABC Blocks.

Mayfair ABC Picture Blocks by Mayfair Novelty Co., c. 1900. Paper litho on wood blocks. Box measures 6.25" x 3.75". Each block measures 1.5" x 1.5". $65-85.

Blocks to Mayfair ABC Picture Blocks.

Acrobat Figure by Crandall, c. 1890. Paper litho on wood. All pieces are connecting. $100-150.

The Embossing Company's Toy Blocks, c. 1890. #7. Paper litho box top with embossed wood alphabet blocks. 4" x 4". $75-125.

Alphabet Figure Blocks by Schoenhut, c. 1900. Paper litho on wood die-cut blocks. One side features a child figure and the other an animal. 2" x 5" tall. $125-225 set.

Opposite side to Alphabet Figure Blocks.

Dr. Drake Picture Blocks, c. 1890. Paper litho on wood, two sided story blocks. 4" x 2.5". $10-22 each.

Fireman Picture Blocks, c. 1890. Paper litho on wood blocks. Alphabet displayed on sides. 2.5" x 4". $45-65 set.

Alphabet and Dominoes Blocks, c. 1890. Paper litho on wood. 1.5" x 1". $45-65 set.

Noah's Dominoes by Crandall's of Montrose, c. 1880. Paper litho on wood. Wood box measures 4.75" x 15". Each domino measures 2" x 4". (Too rare to determine a value.)

Animal dominoes to Noah's Dominoes.

Indian Alphabet Blocks, c. 1890. Paper litho on wood. Reverse side shows American soldiers. 3" x 1.25". $10-25 each.

Alphabet Picture Blocks, c. 1890. Paper litho on wood. Pictures on both sides. 4" x 2.5". $15-25 each.

Block "E" for elephant to Alphabet Picture Blocks.

Block "R" for rocking horse to Alphabet Picture Blocks.

Block "X" stands for 10 and "Y" for Youth to Alphabet Picture Blocks.

Block "S" is for Swing to Alphabet Picture Blocks.

Miss Pussy Cat's School and Unruly Tommie Picture Blocks, c. 1890. Most likely made by McLoughlin Bros. Two-sided paper litho blocks tell a story. Reverse side shows the story of Cousin Dash. 4" x 2.5". $15-25 each.

Close up of block, "Poor Tommie is at last disgraced."

Close up of block, "Mischievous Tommie Inks Miss Pussy Cat's Dress."

Cousin Dash Picture Blocks, c. 1890. Two-sided paper litho blocks tell a story. Reverse side shows story of "Miss Pussy and Tommie." 4" x 2.5". $15-25 each.

Close up of block: "This is our cousin Dash, 'He's welcome,' keeper said."

Close up of block: "Weary and Wet, he leads him home."

The Unequalled A.B.C. and Building Blocks by McLoughlin Bros., 1889. Paper litho on wood blocks. One side is a picture puzzle and the other are alphabet blocks. They feature a very rare triangle shaped top. Only known example. 14" x 12.5". (Too rare to assign a value.)

Train picture blocks from The Unequalled A.B.C. and Building Blocks.

Circus picture blocks from The Unequalled A.B.C. and Building Blocks.

Barnum Museum picture blocks from The Unequalled A.B.C. and Building Blocks.

Robber Kitten and The Three Little Kittens Picture Cubes by McLoughlin Bros., c. 1900. Six-sided paper litho on wood blocks. One side shows the alphabet, the others make five different picture puzzles. Box measures 8.25" x 10.75". Each block measures 2.5". $900-1,300 boxed set.

One picture block scene from Robber Kitten and The Three Little Kittens.

Picture Cubes by McLoughlin Bros., c. 1900. #6570. Nine, six-sided paper litho on wood picture blocks. 8" x 8". $300-600 boxed set.

Wolf and Lamb scene from Picture Cubes.

Bull and Frog scene from Picture Cubes.

Cat pickpocket scene from Picture Cubes.

Child's Picture Block Puzzle, c. 1890. Paper litho on wood box. Twenty, six-sided paper litho on wood blocks. Includes six lithographed sheets with pictures of each puzzle to be used as a guide. 7.5" x 9". $150-300 boxed set.

Boat and swan scene from Child's Picture Block Puzzle.

Baby in a basket scene from Child's Picture Block Puzzle.

Picture Block Puzzle, c. 1890. Nine paper litho on wood, six-sided blocks make six pictures. Comes with six lithographed sheets with pictures of each puzzle to be used as a guide. 4.5" x 4.5". $125-225 boxed set.

Frogs fishing from Picture Block Puzzle.

Six-sided Puzzle Blocks, c. 1890. Twelve large, six-sided paper litho on wood blocks. 6.5" x 6.5". $250-350 set.

Goat pulling sleigh from six-sided Puzzle Blocks.

Set of five Nesting Blocks, c. 1890. Paper litho on wood . Five sides with pictures. The largest block measures 3" x 3.5". $150-300 set.

Stacked blocks in Set of 5 Nesting Blocks.

Set of four Alphabet Nesting Blocks, c. 1890. Paper litho on wood. Four sides with pictures, top shows alphabet letters. 15" high. $100-150 set.

Another view of four Alphabet Nesting Blocks.

Close up of block from set of four Stacking Blocks below.

Set of four Stacking Blocks. Made in Germany, c. 1890. Paper litho on wood. Five-sided pictures. Largest block is 3" x 3". $75-125 set.

Six Stacking Picture Blocks. Made in England, c. 1900. Paper litho on wood. 17" tall when stacked. $100-175 set.

Alphabet Nesting Blocks, c. 1900. Paper litho on wood. Pictures on five sides. Partial set. $150-350 set.

Another view of the Alphabet Nesting Blocks.

Set of five Stacking Blocks. c. 1890. Paper litho on wood. Pictures on five sides. 15" tall when stacked. $120-135 set.

Three Large Stacking Blocks, c. 1900. Paper litho on cardboard. Four-sided pictures with alphabet on top. $50-100 set.

Anther view of the three Large Stacking Blocks above left.

Four Alphabet Stacking Blocks, c. 1890. Paper litho on wood. $45-75 set.

Three Alphabet Stacking Blocks, c. 1900. Most likely German. Paper litho on wood. $40-65 set.

Modes of Travel Cut-Up. Published by Parker Bros., c. 1880. Six sliced spelling puzzles. Paper litho on cardboard. 9.25" x 7.25" *Paul Fink Collection.* $125-250.

Speed — on the Lightning Express.

Quick — can you ride horse back?

Noise — on the Tally Ho.

Large — Across The Ocean.

Swift — Pleasure on the Water.

Above — Up in a Balloon.

Criss Cross Spelling Slips. Goat and Sheep by McLoughlin Bros., c. 1890. Two large strip puzzles. 11" x 11". $100-150.

Home Scroll Puzzle by McLoughlin Bros., 1897. Paper litho on cardboard jigsaw puzzle that is the same as the box cover. 7.25" x 9.25" $75-125.

Goats are very good climbers. Fine leather is made of the skin of the Goat.

From the Sheep men shear wool. The mutton we eat also comes from Sheep.

Natural History by McLoughlin Bros., 1896. Five paper litho on cardboard sliced puzzles. Puzzles include a tiger, horse, ox and cow, hippopotamus and giraffe. 8" x 9". $100-150.

Ride — Can you ride a horse?

Swim — The hippopotamus can swim.

Little Bo-Peep's Picture Puzzles by E.I. Horsman, 1897. Six paper litho on cardboard sliced puzzles with exceptional lithography. 8" x 8". $75-150.

This cruel frog with plumes of red — poor little mouse's blood has shed.

The ugly wolf with Granny's clothes was covered almost up to his nose.

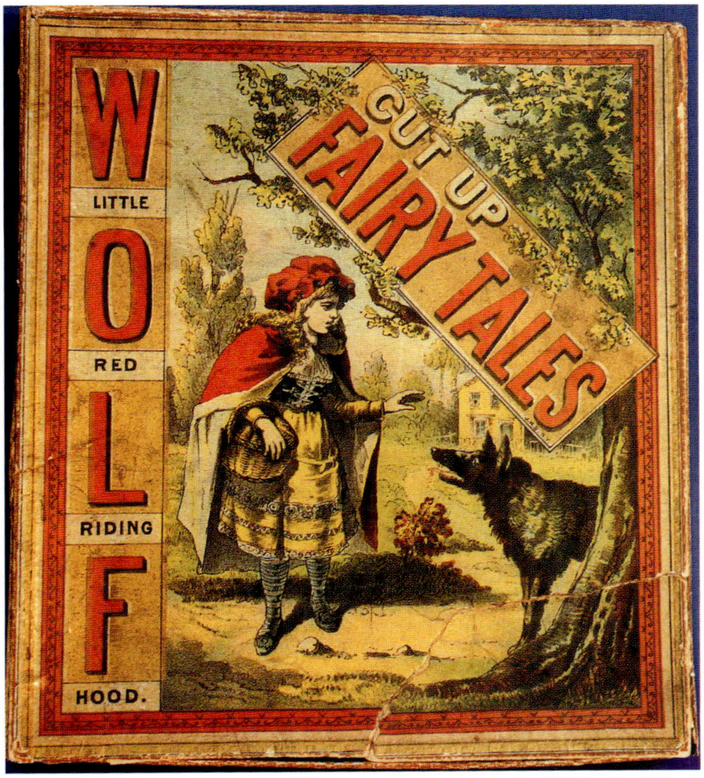

Cut Up Fairy Tales by J.H. Singer, c. 1880. Ten exquisitely lithographed sliced puzzles. 8" x 10". $150-250.

Doom. Jack the Giant Killer.

Love. Did the Beast harm Beauty?

Obey. Can you tell me who this is?

Move. How did she get away?

Criss Cross Spelling Slips by McLoughlin Bros., c. 1885. Set One. Six paper litho on cardboard spelling puzzles. 9" x 10". $95-145.

Boat, c. 1900. Four piece sliced paper litho on cardboard puzzle. 6" x 7". $15-20.

Yacht, c. 1890. Four piece sliced paper litho on cardboard puzzle. 6" x 7". $15-20.

Coach by McLoughlin Bros., c. 1890. Ten piece paper litho on cardboard strip puzzle. 11" x 14". $20-30.

As You Like It Blocks by Ives, c. 1880. Five paper litho on wood puzzles with interchangeable parts. 9.5" x 6.5". $1,500-2,500.

Man with spectacle from As You Like It Blocks.

Man with cap and bow tie from As You Like It Blocks.

Uncle Sam from As You Like it Blocks

Old spinster from As You Like It Blocks

Black man from As You Like it Blocks.

Little Worker's Scroll Puzzle by Milton Bradley, c. 1890. Two paper litho on cardboard jigsaw puzzles. 10" x 7.25". $60-100.

Boy pushing cart from Little Worker' s Scroll Puzzle.

Girl feeding pets from Little Worker's Scroll Puzzle.

Sectional Animals by Milton Bradley, c. 1900. Six paper litho on cardboard jigsaw puzzles. 9.5" x 7". $95-125.

Bear.

Buffalo.

Tiger.

Farmyard Puzzle Box by Milton Bradley, c. 1900. Paper litho on cardboard jigsaw puzzle. 13" x 9.25". $65-135.

Children with donkey from Farmyard Puzzle Box.

Fire Engine Puzzle, c. 1890. Paper litho on cardboard jigsaw puzzle. 7" x 12". $125-250.

Picture Puzzle by McLoughlin Bros., c. 1890. Paper litho on cardboard jigsaw puzzle of a ship. 9" x 12". $125-250.

Three in One Puzzle Box by Milton Bradley, c. 1890. Three large paper litho on cardboard puzzles. 15" x 20". $150-250.

Train puzzle from Three in One Puzzle Box.

Fire Engine puzzle from Three in One Puzzle Box.

Ship puzzle from Three in One Puzzle Box.

Circus Puzzle by Milton Bradley, c. 1890. Large paper litho on cardboard jigsaw puzzle. 11" x 11". $75-125.

Puzzle from Circus puzzle.

Dissected Map of the United States by McLoughlin Bros., c. 1890. Paper litho on cardboard jigsaw puzzle map. $75-150.

Dissected Map of the United States by Milton Bradley, c. 1900. Paper litho jigsaw puzzle of the U.S. 10" x 11.75". $40-60.

Clemens' Silent Teacher by Rev. E.J. Clemens, c. 1890. Dissected map of the United States. Paper litho on wood jigsaw puzzle. Paint advertisement for Sherwin-Williams on the back of the puzzle. 9" x 6.5". $75-125.

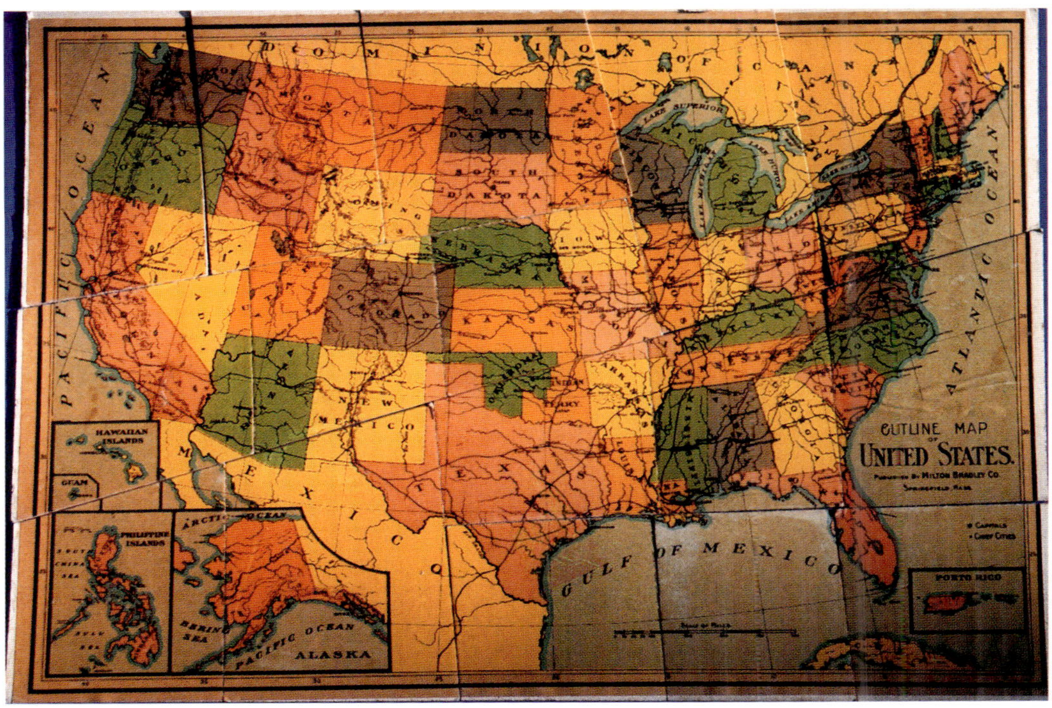

Map from Dissected Map of the United States, above left.

Chapter Four

Dollhouses, Playsets, and Paper Figures

Imagination was the most important commodity a child of the late nineteenth century could own. Toy manufacturers offered beautiful paper lithographed doll houses, cardboard villages, toy soldiers, and dolls for the children to play with in the nursery.

Paper litho villages were beautifully crafted in the late 1880s by The McLoughlin Brothers and Milton Bradley. Each set came in a box complete with buildings, scenery, paper figures, and a play mat to use as a base. The lithography and attention to detail is quite amazing when you consider the small scale of these toys. The average building measured about 4 inches wide by 6 inches high. There were many different sets with many different buildings. Each building was made with litho paper over heavy paperboard and would be sold flat in the box. You would then simply open up the four sided structure and attach a separate roof, or in some cases a roof and a chimney, or a clock or bell tower. "The Pretty Village," by McLoughlin Brothers, was a commercial success because we know that it was in production in different versions for over twenty-five years. "Bradley' s Toy Village" by Milton Bradley was similar to the "Pretty Village," however the buildings were printed in slightly larger scale and the paper quality varied from set to set.

Paper litho dollhouses came in two versions. One was lithographed paper over wood in the Bliss style, the other was lithographed paper on heavy paperboard in the McLoughlin style. Bliss dollhouses were among the most ornately decorated houses offered by American toy manufacturers in the late nineteenth century. Most of them were designed in the late Victorian style with lavishly colored lithographed windows, porches, and columns. The interiors were occasionally decorated in lithographed detail, sporting fireplaces, furniture, pictures, etc. Most Bliss dollhouses were marked with the R. Bliss name. W.S. Reed produced dollhouses similar to those of Bliss. Reed usually used windows that had rounded tops and doors with pediment crowns. Germany also imported paper litho over wood dollhouses to America. The German houses usually had a painted, dark blue roof and are commonly called "blue roof" dollhouses. R. Bliss also produced boxed sets of beautifully detailed paper litho over wood dollhouse furniture. These boxed sets are very hard to come by, however separate pieces of furniture are easy to find.

The McLoughlin Brothers started making dollhouses in 1875. Most of their dollhouses were folding versions made of cardboard. The McLoughlin dollhouses were decorated with unbelievable detail inside and out. The company also produced two sizes of paper litho furniture for their houses. McLoughlin furniture came with small accessories that could be cut out, including pictures with mottoes, china, pets, rugs, and so on.

Paper soldiers, paper dollhouse dolls, paper circus figures, and animals were also very popular at the end of the nineteenth century. Paper soldiers were most prominently produced by The McLoughlin Brothers, Milton Bradley, and Raphael Tuck . They consisted of paper litho over cardboard and usually were attached or could be slipped into small wooden stands. Paper dollhouse dolls and early advertising dolls were the natural accompaniment to the paper litho dollhouse. Along with using color lithography to print trade cards, turn of the century printers also made advertising dolls to sell to companies to give away as advertising premiums. In the late 1800s Lion Coffee made dolls and accessories to coordinate with popular fairy tales and nursery rhymes. The exquisite printing process of chromolithography was once more used to sell everyday products.

The New Pretty Village Engine House Set by McLoughlin Bros., 1897. Box measures 15" x 21". $200-350.

Buildings and accessories included in The New Pretty Village Engine House Set. Playset features six paper litho on cardboard collapsible buildings with folding roofs. Also included are paper play mat, paper litho characters, animals, vehicles, and scenery.

The Volunteer E. Co. No. 1 by McLoughlin Bros., 1897. #8. Paper litho on cardboard fire house. 3.75" x 6". $25-35.

Wheelwright and Blacksmith by McLoughlin Bros., 1897. # 17. Paper litho on cardboard Blacksmith building. 3" x 5". $20-25.

Post Office, Groceries and Drygoods by McLoughlin Bros., 1897. #3. Paper litho on cardboard store. 5.5" x 5". $20-30.

Rip Van Winkle House by McLoughlin Bros., 1897. #16. Paper litho on cardboard house. 4" x 4". $20-30.

The New Pretty Village Church Set by McLoughlin Bros., 1897. Box measures 8.5" x 11.75". $150-225.

Play set from The New Pretty Village Church Set. Set includes four buildings, people, animals, vehicles, scenery, and play mat.

Three Shutter House by McLoughlin Bros., 1897. #4. Paper litho on cardboard house. Shows child playing with hoop. 5" x 5". $25-30.

M.A. Flower Florist by McLoughlin Bros., 1897. #6. Paper litho on cardboard building. 5" x 3". $20-30.

Two Awning House by McLoughlin Bros., 1897. #13. Paper litho on cardboard house. Shows four children playing blind man' s bluff. 5.5" x 5.5". $25-35.

Schoolhouse by McLoughlin Bros., 1897. #1. Paper litho on cardboard. 4.5" x 5". $30-40.

The New Pretty Village by McLoughlin Bros., c. 1900. Five building set. Box measures 11.5" x 8.5". $150-250.

Playset from The New Pretty Village. Five building set with play mat, die-cut figures, and scenery.

Balcony House by McLoughlin Bros., 1900. Paper litho on cardboard house. 6" x 6". $15-25.

Chicken Coop by McLoughlin Bros., c. 1900. Paper litho on cardboard building. 3.75 x 5.5". $15-25.

Bradley's Toy Village by Milton Bradley, c. 1890. Four building paper litho on cardboard set. Milton Bradley had larger scale tiles on their roofs than the McLoughlin Bros. sets. 15.5" x 11". $150-250.

Tiny Tots Hotel by Milton Bradley, c. 1890. Paper litho on cardboard hotel. 6.25" x 6.5". $15-20.

Blue Shuttered House by Milton Bradley, c. 1890. #1. Paper litho on cardboard house that features children walking on stilts. 5" x 6.25". $15-20.

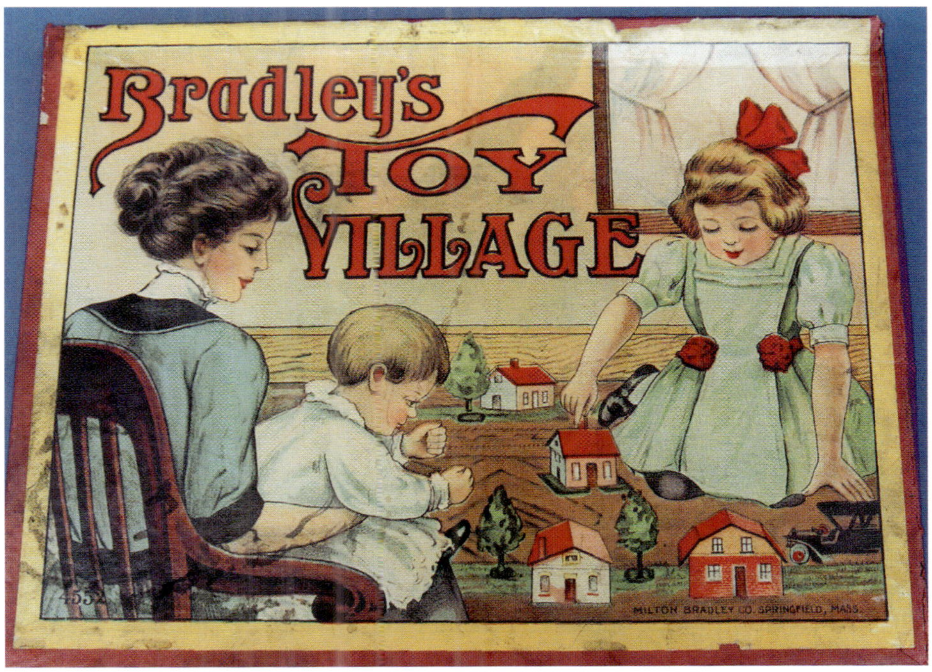

Bradley's Toy Village by Milton Bradley, c. 1890. Six building set. Box measures 7.5" x 9.25". $125-200.

Playset from Bradley's Toy Village. Six building, paper litho on cardboard set with mat and accessories.

Fire Department House by Milton Bradley, c. 1890. Paper litho on cardboard. 6" x 7.5". $15-20.

Bradley's Boat House by Milton Bradley, c. 1890. #4. Paper litho on cardboard building. 3" x 5". $20-30.

Village School by Milton Bradley, c. 1890. #7. Paper litho on cardboard school. 3.75" x 5.5". $20-30.

Uncle Tom's Cabin by Milton Bradley, c. 1890. #16. Paper litho on cardboard cabin. 3" x 5". $20-30.

Red Riding Hood Cottage, c. 1890. Paper litho on cardboard with seashell roof and base. 6" x 4" x 5.5". (Too rare to assign a value.)

Grocery Store by R. Bliss, c. 1895. Paper litho on cardboard building. Three lithograph sides with detachable roof and open back. 8" x 11.5". (Too rare to assign a value.)

Horse Stable by R. Bliss, c. 1898. Paper litho on wood stable, pierced metal railings, and five stalls with composition horses on wheeled platforms (horses not original.) 24" x 25". $2,500-3,500.

Side view of Wild Rose Cottage.

Wild Rose Cottage by R. Bliss, c. 1890. Paper litho on wood and cardboard with wood base. Unusual large scale child pictured at front door. Open back dollhouse with small print paper. 13" x 8.5" x 13". Very rare, only known example. (Too rare to assign a value.)

Small Dollhouse by R. Bliss,. c. 1900. Paper litho on wood. Wood porch and open door. Hinged front opens to two rooms. 7" x 9.5" x 5". $600-800.

Side view of Small Dollhouse.

Gutter Dollhouse, c. 1900. Unmarked paper litho on wood dollhouse commonly referred to as a "gutter house," due to the wood gutter on the roof. Hinged front opens to two rooms. 13" x 9.75" x 4.75". $700-800.

Side view of Gutter Dollhouse.

Bay Window House by R. Bliss,. c. 1900. Paper litho on wood house, five windows, and open door. Features wood turned posts and two rooms. 18" x 10.5" x 8.5" *Linda Boltrek Collection*. $1,600-2000.

Italian Style Dollhouse by Whitney S. Reed, c. 1898. Paper litho on wood dollhouse with three open windows. Hinged front opens to two rooms. 11.5" x 7.5" x 7". *Linda Boltrek Collection*. $1,000-1,200.

Log Cabin by R. Bliss, c. 1900. Paper litho on wood roof, door, and windows. One room house. *Linda Boltrek Collection*. $350-450.

Collapsible Dollhouse by McLoughlin Bros., c. 1900. Elaborate paper litho on cardboard, two story dollhouse. Front folds down to display formal gardens. Detachable roof with two chimneys. 12" x 22" x 10". $350-500.

Opposite page:
Top: Dining Room Set by McLoughlin Bros., c. 1900. Paper litho on heavy paper cut-out furniture set. 13" x 8.5". $40-85.

Bottom: Bedroom Set by McLoughlin Bros., c. 1900. Paper litho on heavy paper furniture set. 13" x 8.5". $40-85.

View of window seat in upper floor of Collapsible Dollhouse.

View of downstairs fireplace in Collapsible Dollhouse.

DINING-ROOM SET.
Paper Furniture.
McLoughlin Bro's New-York
DIRECTIONS.
Cut out neatly. Then to bend straight, rule or score the dotted lines with the point of a pin. This dotted line .... score on the front; this one .... score on the back. Make two pin holes as a guide where to score on the back.
LIVE for Those who LOVE YOU
LOUNGE.
HASTE MAKES WASTE

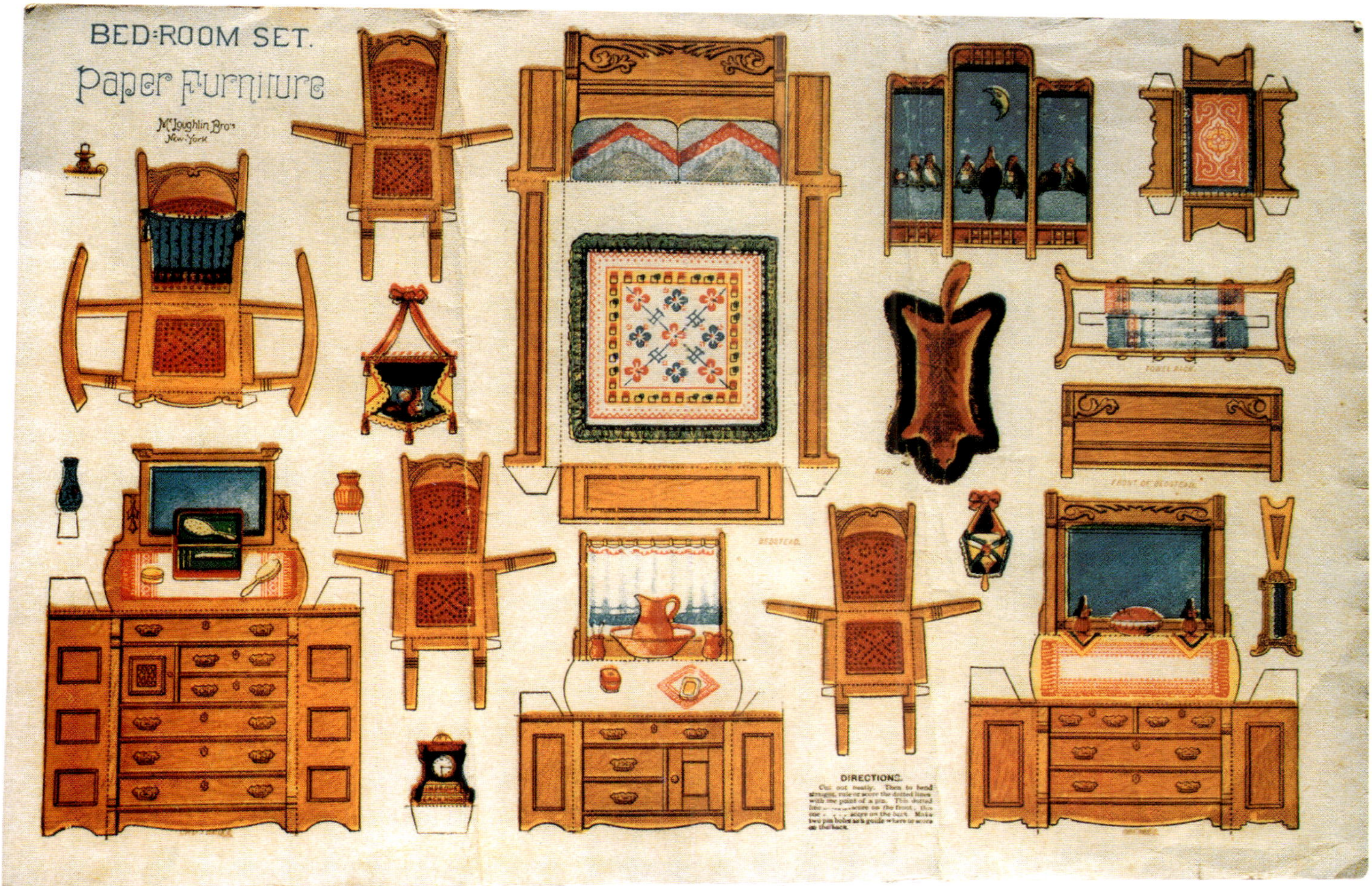
BED-ROOM SET.
Paper Furniture
McLoughlin Bro's New-York
DIRECTIONS.

Floral Dollhouse Furniture, c. 1890. Paper litho on wood, twelve piece set. Possibly French or German. *Linda Boltrek Collection*. $1,000-1,500.

ABC Dollhouse Sofa by R. Bliss, c. 1900. Paper litho on wood. 5" x 3". $200-300.

A.B.C. Furniture by R. Bliss, c. 1900. Eleven piece boxed set. Paper litho on wood dollhouse furniture. Five chairs, sofa, checkerboard table, floor lamp, jardiniere, and piano with bench. Box measures 10.25" x 7.25" (Too rare to determine a value.)

Furniture pieces from A.B.C. Furniture.

Parlor Furniture by R. Bliss, c. 1900. Eight piece boxed set. Paper litho on wood and cardboard dollhouse furniture. Five chairs, sofa, round table, and lamp. 10.25" x 7.25". (Too rare to determine a value.)

Furniture from Parlor Furniture.

Three Piece Lion Cut-out, c. 1900. Paper litho on heavy paper advertising premium. Two children and a buffet. Each figure 5.5" tall. $20-40 set.

Sleeping Beauty by Lion Coffee, c. 1900. Paper litho on heavy paper die-cut stand-up. 5.5" tall. $15-25.

Robinson Crusoe by Lion Coffee, c. 1900. Paper litho on heavy paper stand-up advertising premium. 5.5" tall. $10-15.

Aladdin by Lion Coffee, c. 1900. Paper litho on heavy paper stand-up advertising premium. 5.5" tall. $10-15.

Patriotic Children, 1895. Paper litho on heavy paper stand-ups from Barbour's Irish Flax. Folded outfits fit over paper doll base. Advertising premium. 5" tall. $20-40 set.

Enameline, c. 1900. Paper litho on heavy paper, advertising shoe polish. 5.25". $10-15.

Three Advertising Paper Dolls, 1895. Paper litho on heavy paper advertising premiums from Barbour's Irish Flax Thread. Paper clothing slips over head of paper doll. 4.25" tall. $15-20 each.

Girl with brooms from Three Advertising Paper Dolls.

Girl with flower basket from Three Advertising Paper Dolls.

Two Advertising Paper Dolls, c. 1890. Paper litho on heavy paper advertising premiums from Commonwealth Coffee. Two dolls plus three dresses. $20-40 set.

Paper Doll by McLoughlin Bros., c. 1890. Paper litho on heavy paper with three international dresses. 3" x 8". $30-40.

Paper dresses from Paper Doll.

Two Paper Dolls, c. 1900. Paper litho magazine cut-outs with three outfits. Each doll measures 5.25" tall. $25-35.

Boy with Bicycle, c. 1890. Paper litho on cardboard with slotted wood base. 3" x 4". $20-30.

House That Jack Built Stand-up Playset. Advertising premium from J. C. Ayer & Co., c. 1890. Paper litho on heavy paper. House is 5.5" tall. $35-50.

Circus Stand-ups, c. 1890. Paper litho on heavy paper. Marked "patent applied for." Twelve piece set. 2.5" x 3" each. $30-60 set.

Toy Soldiers by McLoughlin Bros., c. 1890. Paper litho on cardboard soldiers with wood bases. 2" x 7.5". $10-15 each.

Two Toy Soldiers by McLoughlin Bros., c. 1890. Paper litho on cardboard with wood stand. 5" tall. $10-15 each.

Boxed Soldier Set by McLoughlin Bros., c. 1900. Fourteen paper litho on cardboard soldiers. Box measures 13.25" x 11.5". Soldiers are 6.5" tall. $60-100.

Chapter Five

Children's Books

The late Victorian era was overflowing with beautifully illustrated, color lithographed children's books. These books began moving away from teaching lessons and on to entertainment. For the first time, color illustrations were widely used to help young readers visualize the characters and the actions of the stories and poems. Many of the chromolithographic illustrations were works of art in their own right. In the late 1800s, book publishers became more creative and showy.

Publishers like The McLoughlin Brothers, Raphael Tuck, and E.P. Dutton (who used the artwork of the famous Ernest Nister of Germany) introduced books with moveable pictures, die-cut books, 3-D pictures, and books that would open into theaters. Books became a Christmas staple, one that satisfied parents who wanted to give something to their youngsters that would educate and entertain at the same time.

Nursery rhyme books were brought to America by the printer Raphael Tuck & Sons of London. The Tucks did so well in the United States that they opened up an office in New York City. Their "Father Tuck" line of books became runaway bestsellers in America. New York's McLoughlin Brothers became the largest seller of children's literature in the last quarter of the nineteenth century. Their trademark was the "Cinderella' s series" and their stories were drawn from German and English sources.

Unlike lithographed toys and games, many book illustrators signed their work. The main artists of this period were Palmer Cox, Kate Greenaway, Randolph

Left: Aunt Louisa Series, *World-Wide Fables*. Published by McLoughlin Bros., c. 1880. Six full page chromolithographs. 9" x 10.5". $40-60.

Right: Page from *World-Wide Fables*.

Caldecott, and Walter Crane. Some of the best know artists of the time took their hand to illustrate children's books. Also popular during this period were advertising booklets. Many companies would give out nursery rhyme books as premiums to promote their products.

Painting books were also important in the late Victorian age. Children were urged to use paints and a brush even though the crayon had been in use since the eighteenth century. Victorian children were taught to learn through imitation. Painting books often contained beautiful, full page chromolithography and are usually overlooked by children's book collectors.

Collecting chromolithograph children's books is still a very affordable pastime. Although some of the early McLoughlin books can bring in high prices, bargains can still be found. Some collectors like to concentrate on a specific type of book, such as ABC books, linen books, moveable or transformation books, specific fairy tales like The Three Bears or Little Red Riding Hood, painting books, or specific publishers (e.g., McLoughlin Brothers.)

Whatever the category, late nineteenth century children's books are a pleasure to own and a pleasure to read. Many books tell you a story of what life was like after the civil war, a time when children were to be seen and not heard and when the wolf would sometimes eat Grandma or Little Red Riding Hood to teach children an important lesson.

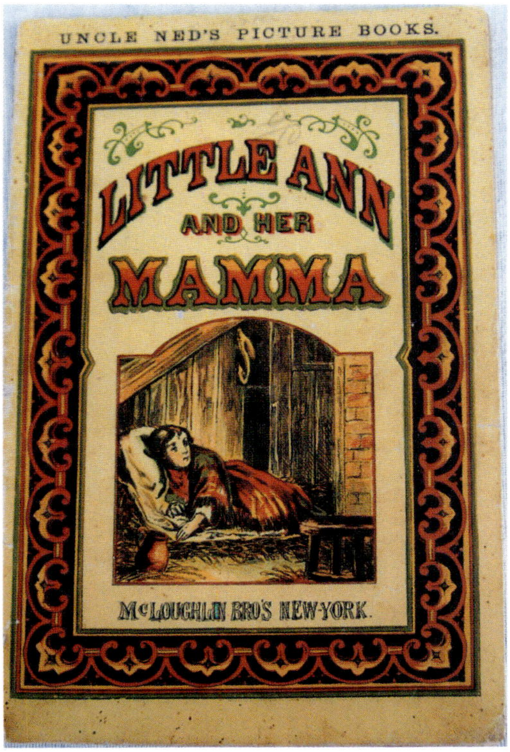

Uncle Ned's Picture Books, *Little Ann and Her Mama*. Published by McLoughlin Bros., c. 1880. Five full color plates. 10.75" x 7". $25-35.

Curly Locks, Little Mother Goose Series. Published by McLoughlin Bros., c. 1900. One color lithograph inside. 7" x 9". $25-40.

Page from *Little Ann and Her Mama*.

Little Pigs, Little Pigs Series. Published by McLoughlin Bros., 1890. Six full page chromolithographs. 8.5" x 10.75". $50-70.

Two pages from *Little Pigs*.

Two pages from *Little Pigs*.

The 3 Little Kittens. Published by McLoughlin Bros., 1890. Hardcover book with fifteen chromolithographs. 12.25" x 10". $60-100.

Page from *3 Little Kittens*.

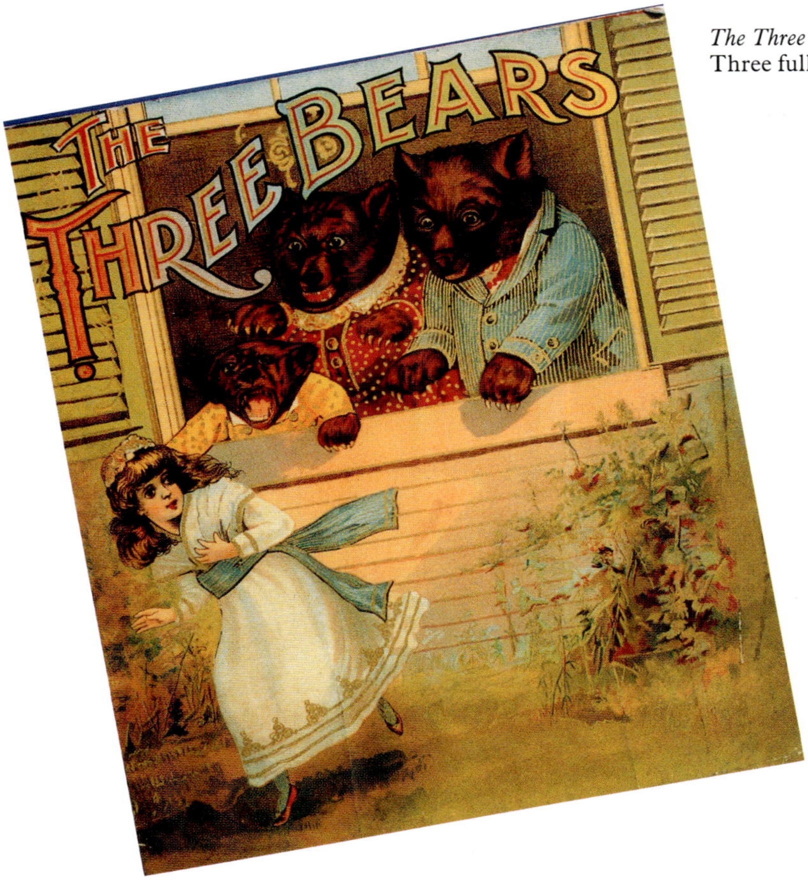

The Three Bears. Published by W.B. Conkley, Chicago, c. 1890. Three full page chromolithographs. 8.5" x 10". $45-70.

Page from *The Three Bears*.

Two page spread from *The Three Bears*.

The Three Bears, Sunshine Series. Published by McLoughlin Bros., 1893. #33. Four chromolithographs. 6.75" x 9". $50-75.

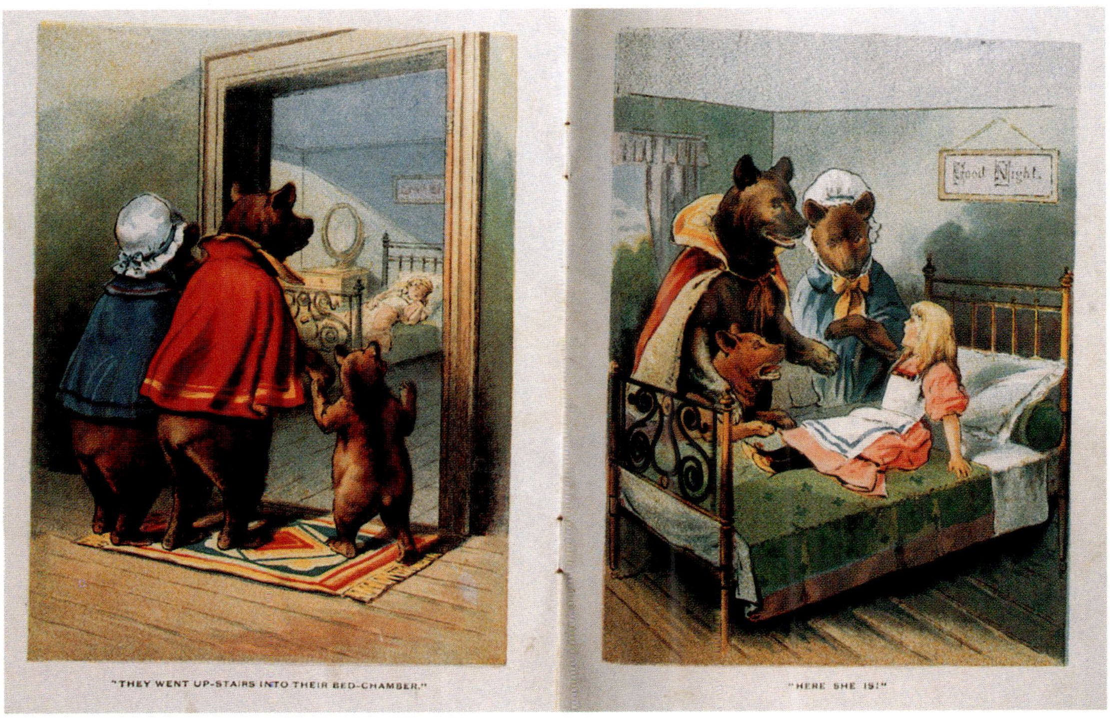

Two page spread from *The Three Bears*, Sunshine Series.

The Three Bears, May Bells Series.
Published by McLoughlin Bros. c. 1890.
Four color plates. 11.5" x 5". $25-40.

Little Red Riding Hood & Other Stories.
Published by Worthington Co., New York.
1887. Hardcover book with black and white
illustrations. 8" x 10". $30-45.

Little Red Riding Hood, Red Rose Series. Published by McLoughlin Bros., 1901. Six full page chromolithographs. 10.5" x 8.5". $60-100.

Two page spread from *Little Red Riding Hood*, Red Rose Series.

Little Red Riding Hood Star Rhymes, c. 1890. Issued by Star Soar as an advertising premium. Sixteen color pictures. 5.5" x 3.25". $20-40.

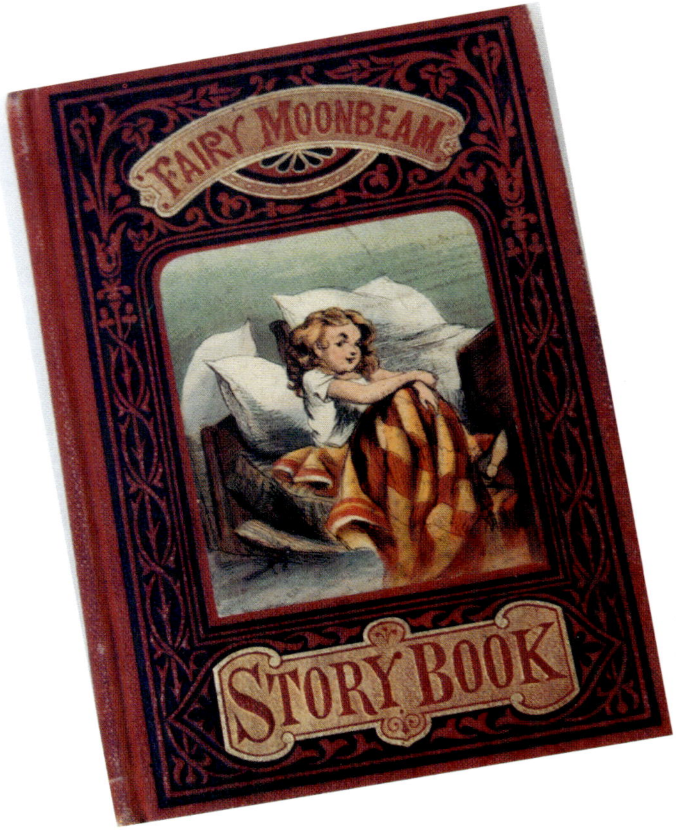

Fairy Moonbeam Story Book, c. 1890. Hardcover book with fifty-four color pictures including Cinderella, Aladdin, Sleeping Beauty etc. 5" x 7". $25-45.

A VERY long while ago, there was a bold, forward little girl, who lived in a far-off country, and the village people called her Silverlocks, because her curly hair was so light and shiny. She was a sad romp, and so full of her pranks, that her parents could never keep her quiet at home. One day, when she had been forbidden to go out, she started off into a wood, to string necklaces of cowslip blossoms, to chase the bees, and to pull down the branches of the wild rose-trees; and she ran about from place to place, until at last she came to a lonely spot, where she saw a

The Three Bears page from *Fairy Moonbeam Story Book*.

Old Dame Trot and Her Comical Cat. Published by McLoughlin Bros., c. 1890. Four chromolithographs. 9.25" x 10.75". $40-60.

The Old Woman and her Pig, Little Pig Series. Published by McLoughlin Bros., 1890. Six full page chromolithographs. 8" x 10.5". *Courtesy of Austern's Antiques*. $40-60.

Double page spread from *The Old Woman and her Pig.*

Jack and the Beanstalk. Published by W.B. Conkley Company, 1897. Three chromolithographs. 8.5" x 10". $20-30.

Frisky the Squirrel, Christmas Eve Series. Published by McLoughlin Bros., 1889. Four chromolithographs. 11" x 8.5". $25-40.

Pets at the Farm. Published by E.P. Dutton & Co., Printed by Ernest Nister in Bavaria. Hardcover book with six chromolithographs. 11.5" x 9". $50-80.

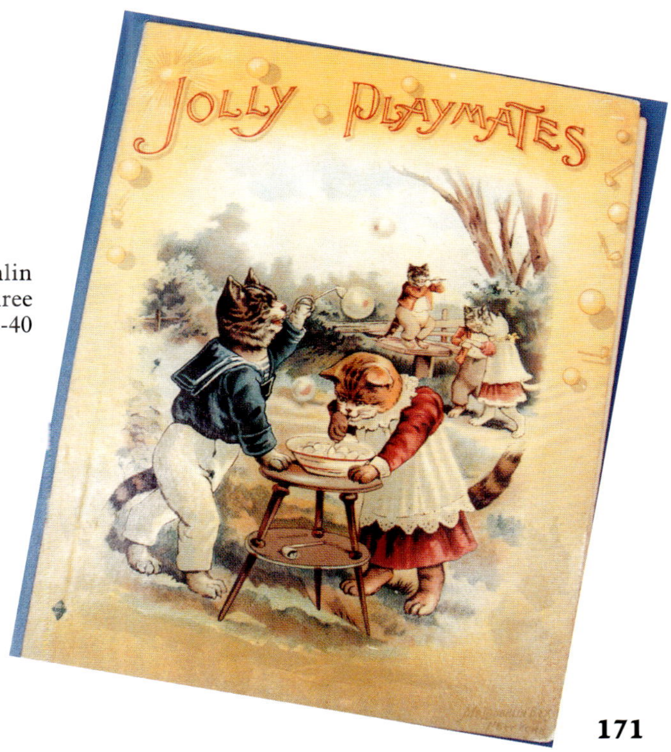

Jolly Playmates. Published by McLoughlin Bros., c. 1890. Hardcover book with three chromolithographs. 8" x 10". $25-40

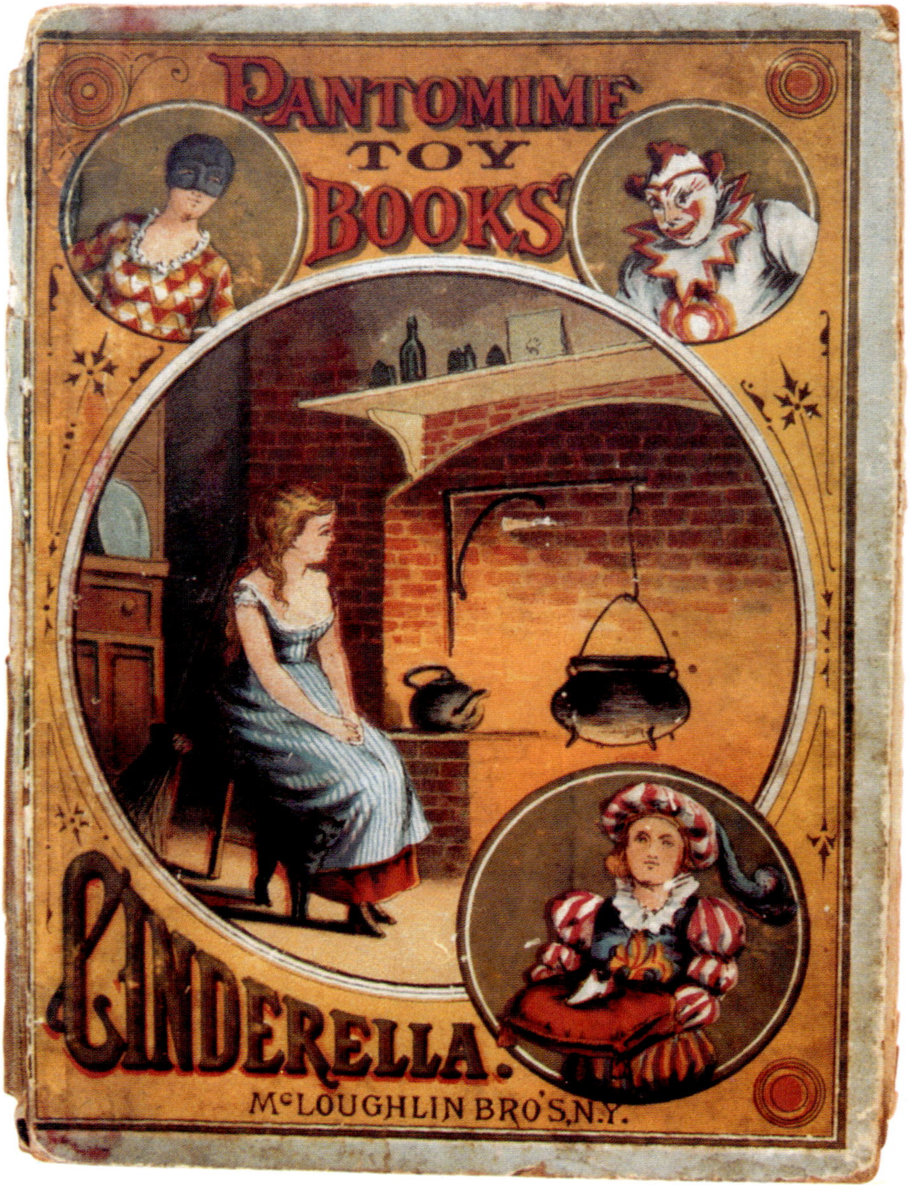

Cinderella Pantomime Toy Books. Published by McLoughlin Bros., c. 1890. Hardcover book opens to a stage with different pages for different scenes. Eleven different flip scenes. 7" x 10". $75-200.

Center page to *Cinderella Pantomime Toy Books*.

Buttercup Farm. Published by E.P. Dutton & Co., Printed by Ernest Nister in Bavaria, c. 1890. Hardcover with twelve chromolithographs. 9" x 11.5". $60-95.

Dissolving Scene by Ernest Nister. Page from a book printed by Ernest Nister. Pull the tab up and the image transforms. $40-50.

Here and There, A Book of Transformation Pictures. Published by E.P. Dutton and Co., Printed by Ernest Nister, c. 1890. Pull tab and the slots move to create a new picture. Five color transformations (ten pictures). 8" x 10". $150-250.

Father Christmas page from *Here and There*.

The House We Live in. Picture Puzzle Toy Books. Published by F. Warne & Co., New York & London, c. 1890. Eight chromolithograph pages. Object of book is to cut out and paste to appropriate scenes. 9" x 10.25". $65-125.

Center page to *The House We Live in*.

Wide World Costume Dolls. Published by McLoughlin Bros., c. 1920. Five full color pages. This book was probably printed by Milton Bradley, who bought McLoughlin Bros. in 1920. 10" x 13". $35-50.

Two page center spread from *Wide World Costume Dolls*.

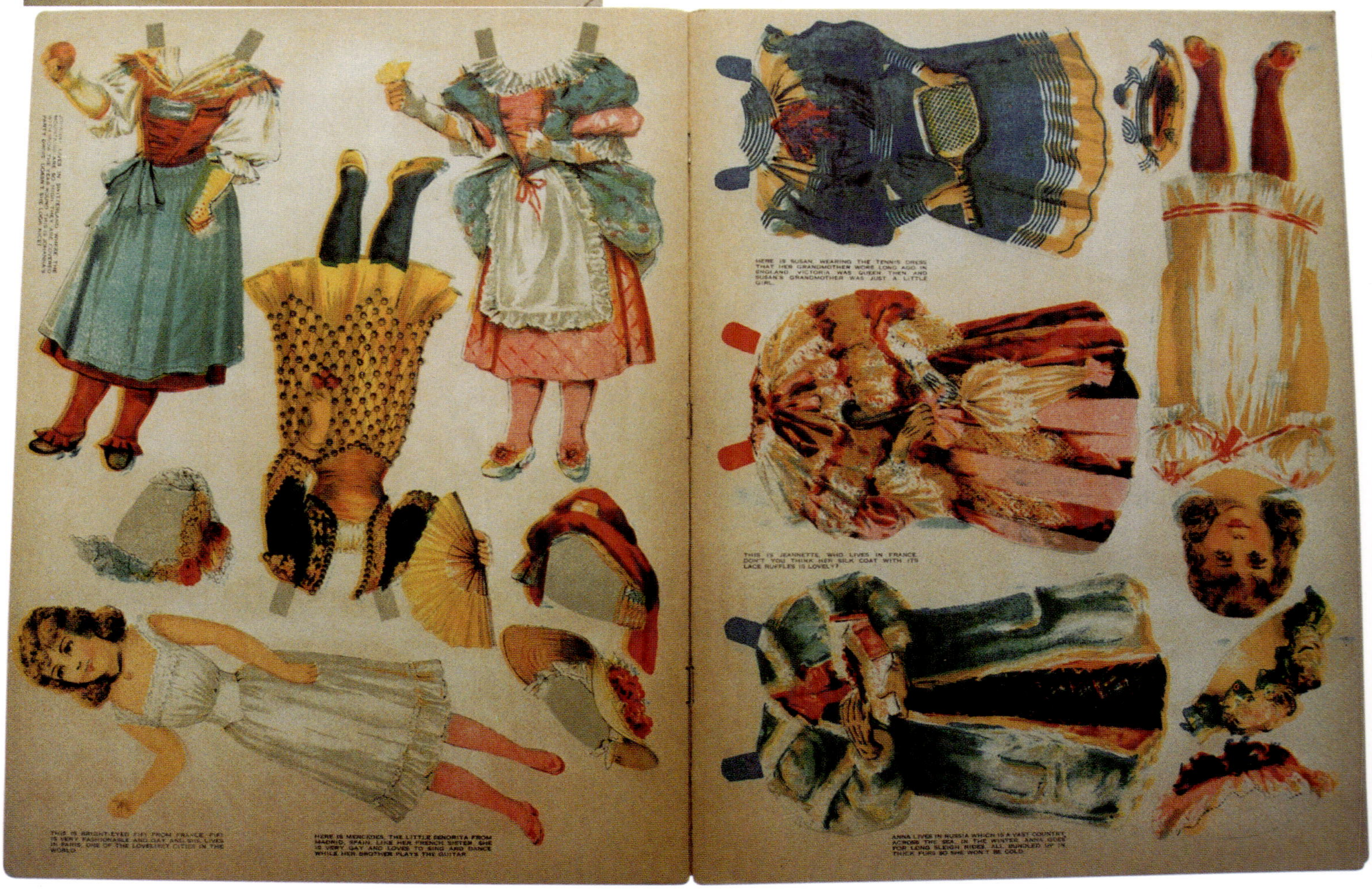

Little Folks Painting Book. Published by McLoughlin Bros., c. 1890. Book No. 3. Eight full page chromolithographs with the opposite page in black and white for coloring. 7" x 7.75". $40-55.

Page from *Litt'e Folks Painting Book*.

Playtime Pictures Painting Book. Father Tuck's Little Artists Series. Published by Raphael Tuck, c. 1890. Four full page chromolithographs. 7.35" x 9.25". $25-40.

Little Folks Playtime Painting Book. Published by McLoughlin Bros., 1898. Six full page chromolithographs. 8.25" x 11". $35-50.

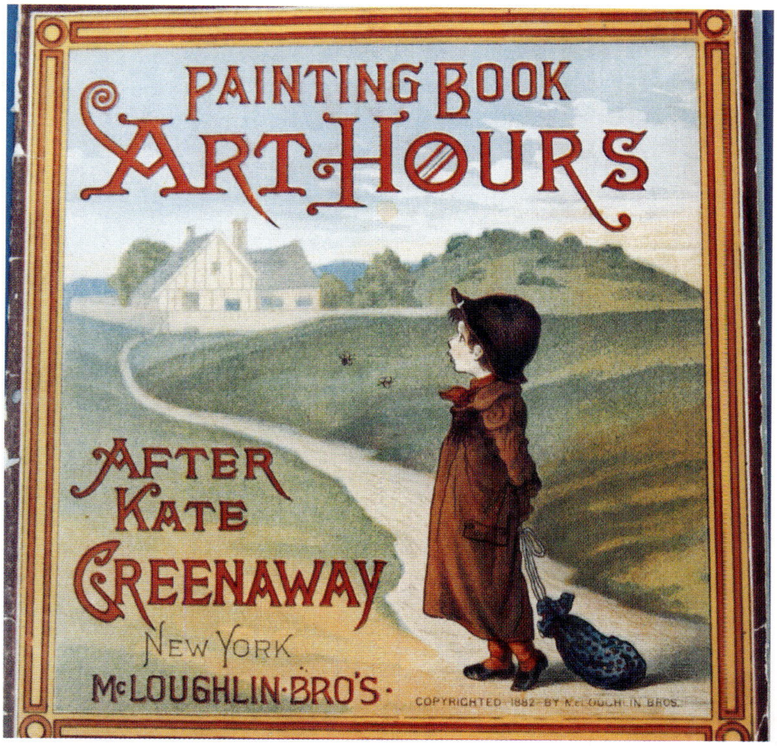

Art Hours Painting Book. After Kate Greenaway. Published by McLoughlin Bros., 1882. Black and white illustrations. 7.25" x 8". $20-30.

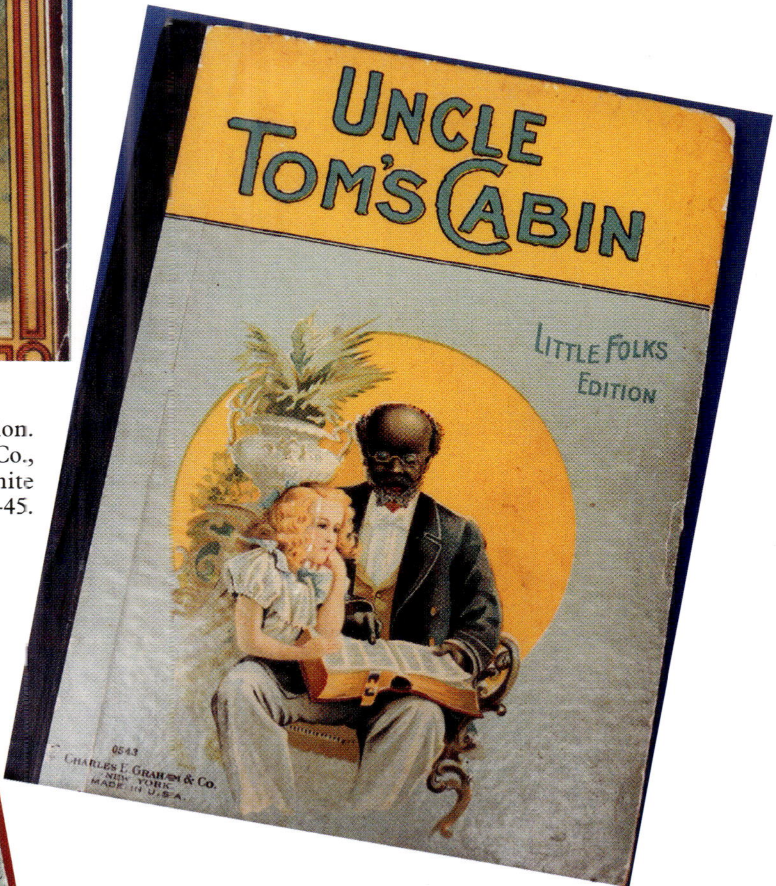

Uncle Tom's Cabin. Little Folks Edition. Published by Charles E. Graham & Co., 1900. Hardcover with black and white illustrations. 10" x 7". $35-45.

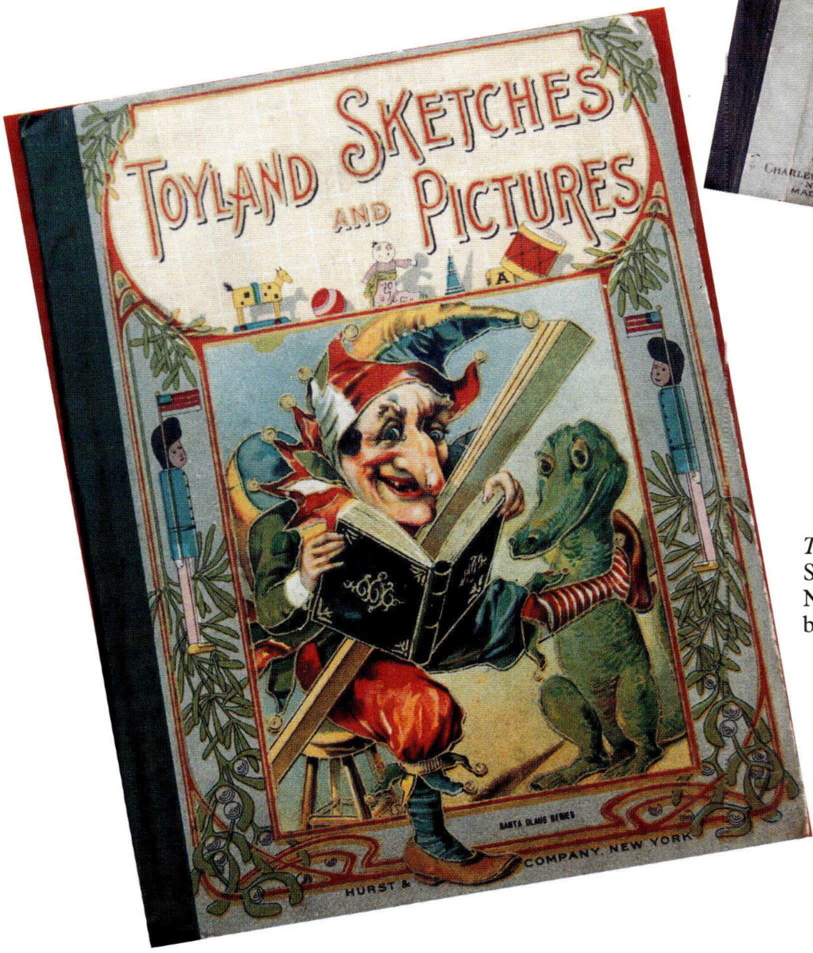

Toyland Sketches and Pictures. Santa Claus Series. Published by Hurst & Company, New York, c. 1890. Hardcover book with black and white pictures. 10" x 7.5". $25-35.

Golden Hours. Mammoth Series. Published by M.A. Donohue & Co., Chicago, c. 1890. Hardcover book with one chromolithograph and black and white illustrations. 7" x 9.75". $25-30.

Little Sunbeams, Nursery Primer Series. Published by M.A. Donohue & Co., c. 1890. #110. Hardcover book with black and white illustrations. 7" x 9.75". $10-25.

Nursery Friends ABC. Father Tuck's "Play & Pleasure" Series A.B.C. Published by Raphael Tuck & Sons, Co. Ltd. Designed at the Studios in England, c. 1890. #1615. Untearable linen with four color pictures. 6.5" x 8.5". $20-40.

Page from *Nursery Friends ABC*.

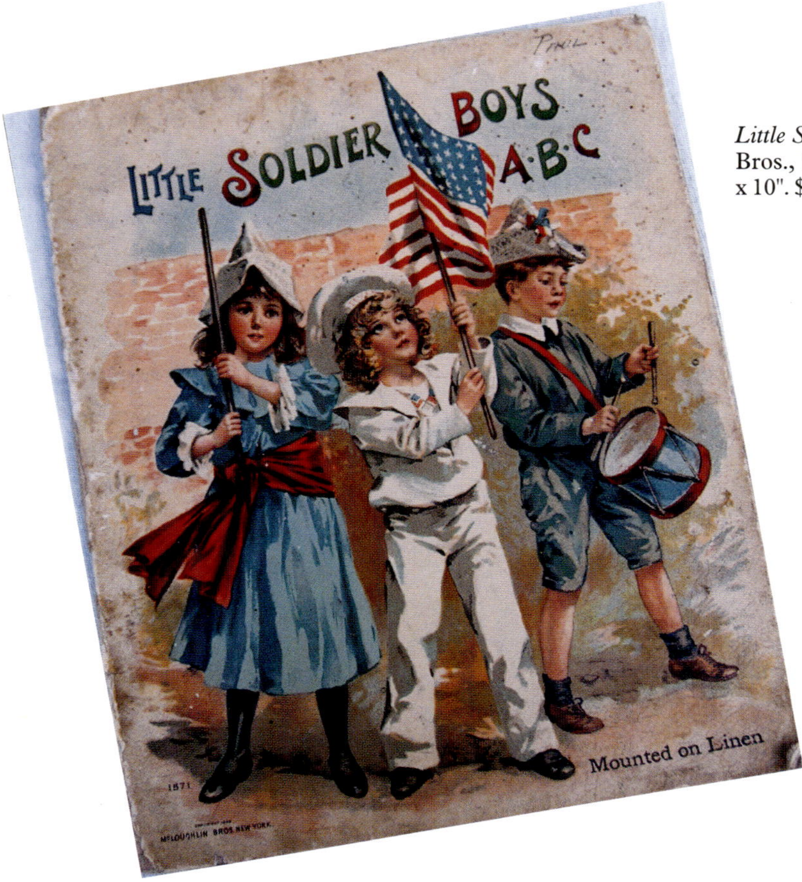

Little Soldier Boys ABC. Published by McLoughlin Bros., 1890. Mounted on linen. Six color pictures. 12" x 10". $25-40.

Pages from *Little Soldier Boys ABC.*

Major's Alphabet. Published by McLoughlin Bros., c. 1885. Early color lithographed alphabet book. 8.5" x 6". $30-50.

Letters U-Z in *Major's Alphabet Book*.

Mammoth Story Book. Published by McLoughlin Bros., 1889. Hardcover book with black and white illustrations. 9" x 10". $20-30.

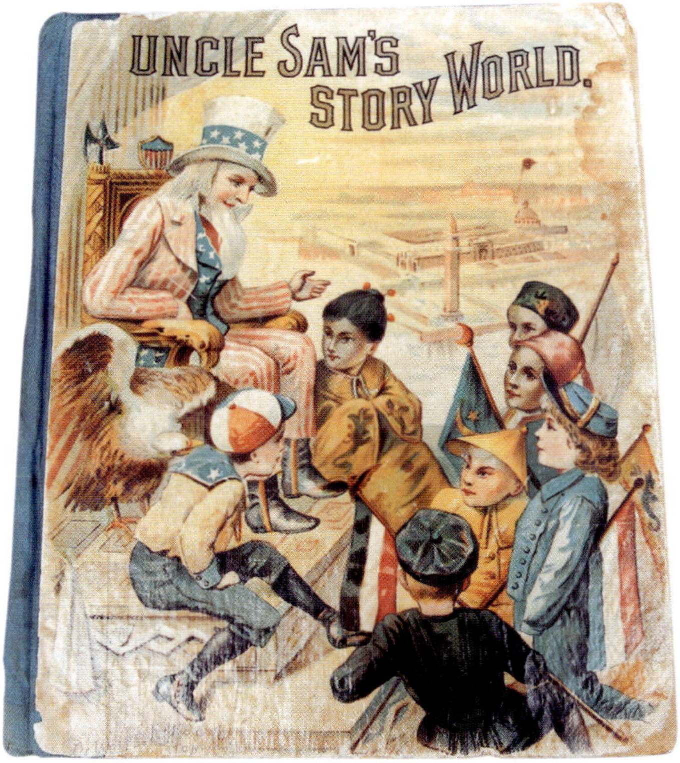

Uncle Sam's Story World. Published by DeWolfe Fiske Co., 1895. One color chromolithograph plus black and white illustrations. 8" x 10". $25-35.

The Night Before Christmas. Published by M.A. Donohue & Co., c. 1890. Hardcover with black and white illustrations. 7.5" x 10". $25-35.

Santa Claus Budget of Good Things. Published by Hurst & Company, New York, c. 1890. Hardcover book with black and white illustrations. 7.25" x 9.75". $25-35.

Nellie's Christmas Eve. Published by McLoughlin Bros., c. 1885. Five color chromolithographs. 9.5" x 10.75". $40-65.

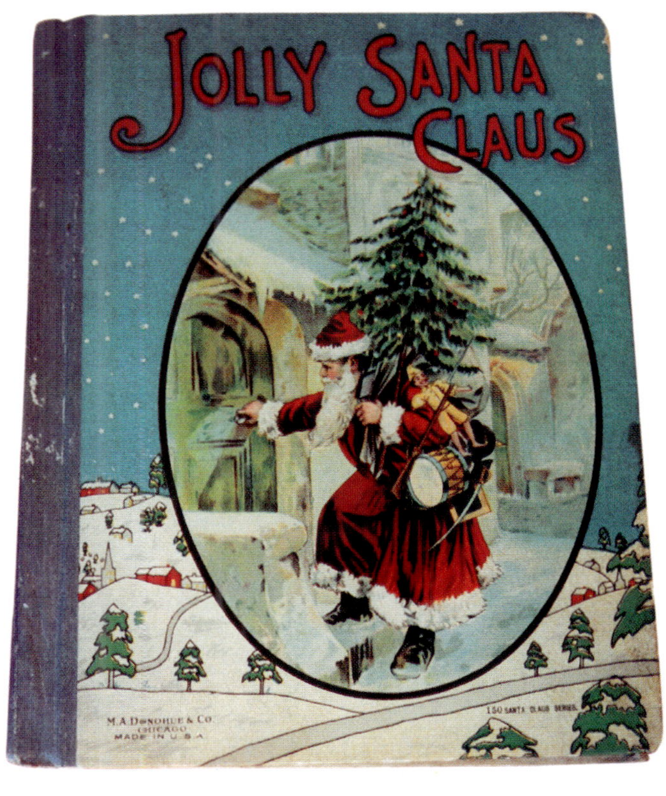

Jolly Santa Claus. Santa Claus Series. Published by M.A. Donohue & Co., c. 1890. Hardcover with black and white illustrations. 7.5" x 10". $35-45.

Chapter Six

Prominent Paper Lithograph Toy Manufacturers

R. Bliss Manufacturing Co.
Pawtucket, Rhode Island

Rufus Bliss started his company in 1832. He began his career as a manufacturer of wood screws and clamps for cabinet and piano makers. Known for their beautifully illustrated paper lithography, Bliss produced games, wood toys, and dollhouses. The earliest known date when Bliss began making paper litho toys was 1871. They produced their most extraordinary paper litho toys during the years 1889 to 1900.

The R. Bliss Company was primarily known for their dollhouses, and their first known dollhouse was produced in 1899. Most of their dollhouses were designed in the late-Victorian style, made of wood and covered in lithographic paper applied in bright and elaborate designs. The dollhouses featured wood carved railings, columns, porches, isinglass windows, and lace curtains. The richly decorated paper lithography was also found on doors, windows, porches, and gables, and sometimes even on the interiors of the houses. Most Bliss dollhouses were identified with the R. Bliss name, a "B" bisected by an anchor, or the R. Bliss Mfg. Co. on a shield.

Milton Bradley & Co.
Springfield, Massachusetts

Milton Bradley was a draftsman and a lithographer and produced his first game in 1860: "The Checkered Game of Life," a board game with high moral overtones (he was the son of a minister). Bradley's early games were meant to be entertainment for soldiers in the Civil War. After the war, he started to make parlor games, was very involved with the Kindergarten movement, and went on to produce many educational games and toys. Milton Bradley believed learning should be fun, a novel idea for his time. Milton Bradley's motto was "Maker of the World's Best Games." Most of Milton Bradley's toys were aimed at children. In 1920, Milton Bradley Co. bought McLoughlin Brothers and reproduced some of their games using the McLoughlin Brothers' images. Milton Bradley died in 1911 but the Milton Bradley Co. was family owned and operated until 1984.

Charles Crandall
Covington and Montrose, Pennsylvania

Charles Crandall was known for developing the first inter-locking tongue and groove paper litho on wood toys. Crandall specialized in building blocks, acrobats, and croquet sets. His toys were inexpensive and promised to keep children busy for hours. Crandall also became known for a game called "Pigs in Clover." In 1889, Selchow & Righter Company became the sole agents for the Crandall Company. When Charles Crandall was eulogized in *The New York Tribune* they said, "In his day he probably furnished more pleasure through toys and puzzles than any other person in the country."

E.P. Dutton
New York, New York

The E.P. Dutton publishing company was best known for its collaboration with Ernest Nister of Nuremberg, Germany. The Dutton firm was founded in 1852 in Boston, Massachusetts, and moved to New York when Edward Dutton bought out his brother's interest in the company. The E.P. Dutton Company was given exclusive rights to distribute and sell Nister books in America. They published many moveable and transformation books under this agreement.

Gibbs Manufacturing Co.
Canton, Ohio

Lewis E. Gibbs was the founder of Gibbs Mfg. Co., in 1884. He began as a producer of farming plows and later added toys to his line in 1886. Gibbs Mfg. Co. is most well known for their paper litho on wood articulated horses that pulled assorted wagons and carts. Gibbs also made advertising toys for different companies.

E.I. Horsman
New York, New York

E.I. Horsman was known mostly for their game "Halma." Many of Horsman's game boards are beautifully lithographed and rival those of McLoughlin Brothers. Horsman was one of the first companies to advertise their toys in magazines, dating back to 1887. At the beginning of the 1900s, E.I. Horsman stopped making games and instead went on to produce children's dolls.

W. & S. B. Ives
Salem, Massachusetts

W & S. B. Ives was established in 1830. They were originally in the book publishing, binding, and selling business at the Old Corner Bookstore. They were one of the first major manufacturers of games in the United States. Their game, The Mansion of Happiness, was introduced in 1843 and was one of the first board games in the United States. Ives was also known for their card games, especially Dr. Busby. Mansion of Happiness and Dr. Busby were both hand colored; an assembly line of artists would paint only one color at a time. In the late 1860s, color lithography took over. By 1898, after the death of the last Ives brother, George S. Parker acquired the rights to the Ives games and reissued them.

I.B. & W
Bridgeport, Connecticut

Ives, Blakeslee & Williams started making toys in the late 1860s. They were known for their quality clockwork toys. They produced only a few games and should not be confused with W. & S.B. Ives. Their toys were marked I.B. & W.

McLoughlin Bros.
New York, New York

The McLoughlin Brothers were the "Kings of Chromolithography" in America. The lithographed illustrations on their toys, games, and books are the most sought after by collectors.

In 1828, John McLoughlin Sr. started his company in New York City as a book publisher. His son, John Jr. expanded the line to hand colored card games. From 1880 until the 1890s, McLoughlin Brothers' factory in Brooklyn had the largest color printing plant in the United States. Milton Bradley bought McLoughlin Brothers in 1920.

Ernest Nister
Nuremberg, Germany

Ernest Nister Company was one of the most prominent publishers and printers of moveable books. Ernest Nister worked in cooperation with E.P. Dutton, who was Nister's representative in the United States. Nister was known for his different types of changeable books. Some books rotated in a circle, some were set up as a pop-up book, and still another type operated on the Venetian blind principle (in which you pulled a tab and the picture would transform into a different view). Nister' s books were very delicate and it is very rare to find one today in excellent condition.

Parker Brothers
Salem, Massachusetts

George S. Parker produced his first game at the young age of sixteen. The game was called "Banking." While Milton Bradley geared their line to children, Parker Brothers went after the adult market as well as the children's. George started his own company in 1883. His brother joined him in 1888 and the firm became Parker Brothers. George invented over one hundred games in his lifetime and his games won the highest award at the World's Colombian Exposition of 1893. He purchased the rights to the games of W. & S. B. Ives and also sold the games of J. H. Singer. Parker Brothers was incorporated in 1901, which helps to date some of their games. The company was family owned until 1968.

W. S. Reed
Leominster, Massachusetts

The Whitney S. Reed Toy company was founded in 1875. They were known for their paper litho on wood toys, which included children's furniture, wooden ships, dollhouses, and other paper on wood toys. The company was sold in 1897 and the name was changed to the Whitney Reed Chair Company, although they continued to make toys.

A. Schoenhut & Company
Philadelphia, Pennsylvania

Albert Schoenhut began manufacturing toy pianos in 1872. The company was also known for framed lithographed cardboard figures animated by a clockwork mechanism. They also made paper litho on wood toys.

J. H. Singer
New York, New York

Jasper Singer was an early producer of paper lithographed games and toys. They were known for their small board games with intricate designs, produced from the 1880s until the 1890s. Toys were also marked J.H.S.N.Y. or J.H.S.

J. W. Spears & Sons
London, England

Founded in 1877, J. W. Spears & Sons were producers of games, theaters, and other paper lithographed toys. They were based in London but also had printing works near Nuremberg, Germany, where chromolithography was at its best.

Raphael Tuck & Sons Ltd.
London, England

Raphael Tuck began his career in 1866 selling prints and frames in a small shop in England. The business thrived and Raphael Tuck & Sons became known as a great worldwide publishing company. Tuck & Sons were known for their children's books, paper toys, paper dolls, and postcards. In the late 1800s, Raphael Tuck had offices in New York, Paris, and Berlin. There is little history of the company available to the collector due to the destruction of the London based plant during World War II. In 1880, the company registered their famous trademark, the "easel and palette" with the inscription "The Worlds Art Service."

United Game Co.
Brooklyn, New York

The only thing known about The United Game Co. was that they were somehow related to Parker Brothers. Their games had the same illustrations as Parker Brothers and some even bore the Parker Brothers' ship logo. The United Game Co. games also included a card advertising Parker games. The company was in business in the early part of the twentieth century.

Frederick Warne & Co.
London, England

Frederick Warne started his book publishing company in 1865 with a New York branch established in 1881. The company was well known for publishing the books of Victorian author-illustrator, Beatrice Potter.

Bibliography

Bajorek, Lagretta Metzger. "America's Early Advertising Paper Dolls." *Antiques Journal*, May 1999, 74-75.

Barlow, Ronald S. *The Great American Antique Toy Bazaar.* El Cajon, California: Windmill Publishing Company, 1998.

Dennis, Lee. *Warman's Antique American Games.* Radnor, Pennsylvania: Wallace-Homestead Book Company, 1991.

Goodfellow, Caroline. *A Collector's Guide to Games and Puzzles.* London, United Kingdom: Quintet Publishing Ltd., 1991.

Hockenberry, Dee. *More Enchanting Friends, Storybook Characters, Toys, and Keepsakes.* Atglen, Pennsylvania: Schiffer Publishing Ltd., 1998.

Howard, Miriam B. *Those Fascinating Paper Dolls.* Toronto, Canada: General Publishing Company, 1981.

Jones, Diane M. and Rosemary. *Collector's Guide To Children's Books.* Paducah, Kentucky: Collector Books, 1997.

Kaonis, Donna C. "The Art of the Game." *The Inside Collector*, September/October 1994, 31-36.

Kaonis, Donna C. "Printed Pastimes: Nineteenth-Century Paper Lithographed Toys." *Collectors' Eye*, November/December 1998, 17-21.

McClinton, Katherine M. *Antiques of American Childhood.* New York, New York: Clarkson N. Potter, Inc., 1970.

O'Brien, Marian Maeve. *The Collector's Guide To Dollhouses and Dollhouse Miniatures.* New York, New York: Hawthorn Books, 1974.

Porter, Catherine. *Miller's Collecting Books.* London, United Kingdom: Reed International Books Ltd., 1995.

Rinker, Harry L. *Antique Trader Books Guide To Games & Puzzles.* Dubuque, Iowa: Antique Trader Books, 1997.

Whitehill, Bruce. *Games, American Boxed Games and Their Makers.* Radnor, Pennsylvania: Wallace-Homestead, 1992.

Whitton, Blair. *Bliss Toys and Dollhouses.* New York, New York, Dover Publications, Inc., 1979.

Whitton, Blair and Margaret. *Collector's Guide to Raphael Tuck & Sons.* Cumberland, Maryland: Hobby House Press, 1991.

Whitton, Blair. *Paper Toys of the World.* Cumberland, Maryland: Hobby House Press, Inc., 1986.

Zillner, Dian, and Patty Cooper. *Antique & Collectible Dollhouses and Their Furniture.* Atglen, Pennsylvania: Schiffer Publishing Ltd., 1998.

Index